FOALS AND YOUNG HORSES

Training and Management for a Well-behaved Horse

Horse Riding and Management Series

5m Publishing

FOALS AND YOUNG HORSES

Training and Management for a Well-behaved Horse

Horse Riding and Management Series

UTE OCHSENBAUER

BEATE SCHMIDTLEIN

Translated by Sue Anderson

5m Publishing

First published 2013

This edition published by 5m publishing 2016, reprinted 2022

Sue Anderson asserts her right to be known as the translator of this work.

Published by
5M Publishing Ltd,
Benchmark House,
8 Smithy Wood Drive,
Sheffield, S35 1QN, UK
Tel: +44 (0) 1234 81 81 80
www.5mpublishing.com

A Catalogue record for this book is available from the British Library

ISBN 978-1-910455-09-8

Book layout by
Keystroke, Station Road, Codsall, Wolverhampton

Printed by
Replika Press Pvt. Ltd, India.

Photos by Horst Streitferdt/Kosmos, Gaby Kärcher, Stefan Lafrentz, Beate Schmidtlein

Illustrations by Esther von Hacht

All of the information and methods in this book have been carefully researched, considered and checked. They do not release horse lovers from their responsibility for their animals or for themselves.

The methods described are used under the reader's own responsibility. The publisher and authors assume no liability for loss, damage or injury to persons or property, or for consequential damages, arising from the use of the equipment and methods presented in this book.

FSC www.fsc.org MIX Paper from responsible sources FSC® C016779

About the authors

Through her training and her work on stud farms, Ute Ochsenbauer also has vast experience of the day-to-day handling of foals and young horses. Ute has kept horses for over thirty years and trains them herself.

As a successful breeder of Trakehners and a fully qualified stud manager, Beate Schmidtlein has a lifetime's experience of rearing foals and training young horses. Over the years Beate has devised a very gentle programme of rearing and weaning for her foals.

Contents

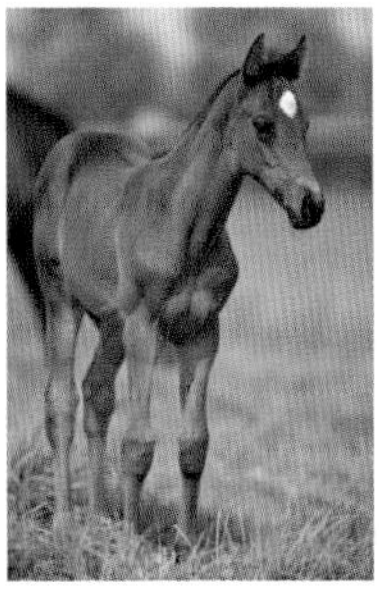

For organisational reasons, all photographs specially commissioned for this book were taken during a summertime photo session – including those for the winter chapters.

Headway charity disclaimer

The books in the 'Horse Riding and Management Series' were originally published in Germany. The photos used in the books are taken from the German books and in some cases show riders not wearing protective helmets. The law on wearing protective headgear differs between countries, states and equestrian disciplines. However, it is strongly advised that all riders wear a properly fitting, CE approved helmet at all times when riding, whether on private or public ground, for all riding activities. The UK Highway Code makes it clear that children under the age of 14 must wear a helmet that complies with the Regulations, and other riders should also follow these requirements. This applies to roads, bridleways, footpaths, cycle paths and other roadways.

5m have partnered with Headway – the brain injury association – on this issue to highlight the importance of children and adults wearing appropriate head protection while riding. The following statement has been provided by Headway:

> Headway, the UK-wide charity that supports people affected by brain injury, strongly advises riders of all ages and abilities to wear hats with straps that meet British standards.
>
> We all think 'it will never happen to me', but the reality is that an accident can happen to anyone at any time, with all riders at risk – regardless of age or experience.
>
> If worn correctly, riding hats are effective at preventing head and brain injuries, the effects of which can be devastating and last a lifetime.
>
> To all horse and pony riders, we simply say: Please, use your head, use a hat.

For further information about brain injury and how Headway helps those affected, or to make a donation to the charity, please visit www.headway.org.uk or call the freephone helpline on 0808 800 2244.

Headway
the brain injury association

Foreword: From child of nature to trained partner

Newborn foals are a 100% natural product. It is only in the first few years and months of their lives that we humans intervene, aiming to refine them into useful assets. As we rear horses, we want to retain their trusting curiosity, friendliness, readiness to learn, natural impulsion and drive, plus all of their other positive traits.

Our job as handlers is to help a foal to develop gradually from an instinct-driven child of nature into a trained, specialised partner in our riding and driving activities. We expect the horses we work with to be cooperative and ready to perform. This is certainly achievable if we can learn from them how to be 'present' and how to master a language that we have long spoken without knowing it: body language. All of the skills that we develop in ourselves or our four-legged partner during the civilising process will serve us well throughout its life – either in our general contact or in our riding partnership. So it is vital to proceed with caution, patience and an eye on the long term!

And as owners and riders, we can also continue to develop ourselves, and to improve our 'horse sense', our capacity for learning and our fairness and fitness. This book is designed to help. Learning is easier when done together!

Learning is easier when done together.

The nursing foal to weaning

Before you decide to breed

AGE
0 to around 8 months

Should I breed my mare? How do I rear my foal? What should it learn as a newborn, or as a yearling or a two-year-old, in order to grow up to be a friendly, easy-to-ride horse that trusts its handler? These are questions that preoccupy many mare owners.

And many horse lovers who dream of having their own horse wonder whether to buy a foal rather than a fully-grown animal. If you breed a foal yourself, you need to be ready to accept what nature gives you. This could be anything – from a champion stallion to a reliable 'happy hacker', from a perfect healthy specimen to a sick or even a dead animal. Even at the weaning stage, a foal's size, character and rideability in later life can still only be guessed at. If you buy an older horse, on the other hand, you can draw fairly accurate conclusions about its character and health.

Inexperience and lack of knowledge often lead to uncertainty about the rearing and training that a young horse should receive. We have heard stories of weanlings that were already being taught lateral gaits, nursing foals that were in scheduled training programmes, separated from their mothers for hours on end, and yearlings that were already being lunged under saddle.

Training too early causes serious, lasting damage to the immature body and mind of a young animal. Horses learn best in an atmosphere in which they feel safe. If their stress levels are too high, they can become traumatised. (For more on stress and optimum conditions for learning, see pp. 64, 65, 67, 130)

On the other hand, we know of many foals that may well be growing up under natural herd conditions but have almost no contact with people. Animals like these are often anxious and defensive during vet or farrier visits, changes of pasture or treatment for injuries. Their owners may then be tempted to use coercion (e.g. restraint, tethering, a twitch), pressure or even violence to allow the vet or farrier to do their job, to change pasture or to treat the injury. This teaches the young animal to associate people with fear and anxiety. It will take quite some time for the animal to develop the deep trust that we want from our horses.

Many difficult behaviours in adult horses have their roots in management or training methods that are horse-unfriendly, unnatural or technically

Rhapsody feels safe under the watchful gaze of her aunt!

Guarding foals is a job often shared between friends or related mares within a herd. Here, Starlight is Rhapsody's aunt.

incorrect. Inexperience and lack of knowledge often lead to misunderstandings and injustices, with potentially serious consequences.

The aim of this book, therefore, is to improve understanding between young horses and their owners. We want to support you in building a sound, trusting relationship with your young horse and to help you rear it in an age-appropriate way. Problems at a later stage can be avoided by striking the right balance between holding back and 'being present'.

We hope to be able to answer many of your questions by giving practical examples and providing plenty of additional information.

DEEP TRUST FROM THE START

At around half past six one morning, I went out to the stables to give our little herd their first hay of the day. We have an open stable, where we close the stall doors for feeding so that all animals, even low-ranking ones, can eat in peace. I am usually greeted by the same sight every morning. The horses are all lying down in their open stalls.

The Haflinger jumps to his feet immediately but only then does he really wake up; the New Forest pony lingers for a moment longer before getting up, and the half-bred casts a soulful eye in my direction and goes back to sleep. He doesn't stir, not even when I put his hay down right in front of him. I can approach him, or even lie against his side and stroke him, if we are both happy with that.

All three of these animals grew up in a bachelor herd of young stallions. The Haflinger had contact with people only when a change of pasture, farrier's visit or veterinary attention (e.g. castration) were required. The New Forest pony had slightly more contact with people. Our half-bred came to us at the age of one having had regular human contact. He has a fundamental trust in humans that is as deeply ingrained as the fundamental distrust shown by our Haflinger, who came to know humans as creatures that would approach him only for particular reasons and then carry out their plans regardless.

Wouldn't we all like to have horses that trust us so implicitly that they stay lying down when we approach? A path of trust, right from day one.

POINTS TO CONSIDER BEFORE BREEDING OR BEFORE BUYING A FOAL

- Can you provide ideal conditions for the pregnant mare?
- Is she living in a peaceful herd, does she get plenty of exercise every day?
- Is she healthy?
- Can you afford to provide medical care for the mare or foal if something goes wrong?
- Can you afford a second horse?
- Can the mare stay in familiar surroundings after she foals?
- If not, where could she raise her foal?
- Does the foal have playmates of its own age?
- Where can the foal go after weaning, to allow it to grow up naturally? This point should be considered separately for colts and fillies, as there are limits to allowing colt foals to run with mares.

If answering these questions has convinced you to breed your mare or to buy a foal, we wish you luck. If you decide to breed, we wish you and your mare good luck in your choice of sire and a trouble-free pregnancy.

The final stages of pregnancy

The average horse pregnancy lasts for around 11 months (332 to 335 days). But anything between 310 and 365 days is possible. The length of pregnancy is influenced by factors such as breed, time of year (of mating and foaling), light levels, weather conditions and food supplies. General condition, illness, stress, being sold during pregnancy, getting used to a different environment and loss of friends are other factors which determine the length and progress of pregnancy.

The mare close to foaling

In the later stages of pregnancy, most mares start to show that they are carrying a foal. Many mares become calmer as soon as they are pregnant, and their facial expression changes.

Sometimes a pregnancy can be identified just from this change of expression, even before the vet has confirmed it.

The pregnant mare feels secure in a peaceful herd of mares.

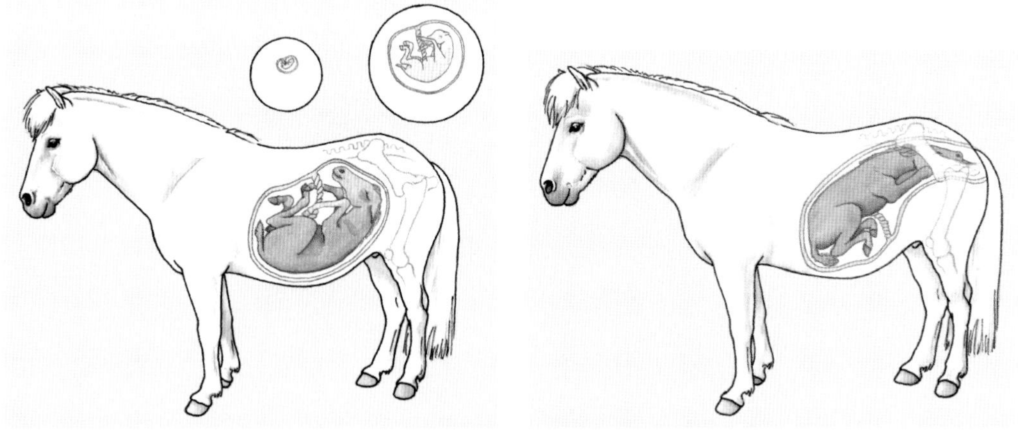

Until the eighth month of pregnancy, the foal grows very slowly. Then space starts to get tight!

Mares are naturally more rectangular in build than stallions. Pregnancy is often spotted late in long-bodied mares, even by experienced breeders. Shorter-bodied mares, on the other hand, often start to 'show' much sooner.

Many horse lovers will have heard tales of mares unexpectedly delivering a foal the day after an examination, competition or particularly strenuous ride. Or they may even have seen it themselves. But it is much more horse-friendly, and much safer, to take good care of your mare at this stage.

In earlier times, when horses were still an essential part of the agricultural economy, mares were worked right up until foaling. The more exercise a mare gets, the easier she will find the birth. Gentle hacks are recommended at every stage of pregnancy.

Strenuous riding, high performance sport, jumping, tight turns, tournaments and the associated stress of travelling could all be stressful for a pregnant mare. In the worst-case scenario, they could even cause her to lose her foal.

Spontaneous abortion can also be caused by falls, bumps or kicks, poor management or feeding, or excessive worming. Herpes virus infection can trigger epidemic abortion. This can be prevented by vaccinating mares appropriately. However, vaccination is controversial and works only if all horses on a yard are treated.

People who know a mare well can often spot a pregnancy immediately in her face.

The needs of the pregnant mare

Horses feel most at ease if there are not too many changes in their herd or their daily routine. Unfamiliar feeding times or a change of stall or pasture is enough to put them on edge. The safety and security that mares in particular experience in a stable herd cannot be overestimated. Ideally, mares in late pregnancy should not be moved from their herd or stable.

All calculations of foaling date are merely estimates. So it is important to observe your mare closely, to keep her healthy and to make sure she has plenty of fresh air, light and daily exercise, ideally on pasture. Our mares foal once the grass starts to grow, providing an ample supply of nutrients. The nights begin to get milder from around mid-April. But we still bring mares inside overnight if they are close to foaling. This enables us to observe them at night and to be on call immediately if there are any problems. However, some mares have failed to appreciate our efforts and have arranged things so as to foal at pasture in the middle of the day!

Caring for the mare as the birth approaches

The peak growth time for the foetus is in the last third of pregnancy. At this stage the foal seems to turn sharply and change position. Some mares feel a bit unwell around this time, which can affect their feeding behaviour. Some mares show particular likes or dislikes in the final weeks of pregnancy. Types of feed that they would normally eat with relish are suddenly rejected. For example, we give an oat/barley/mineral mix as a concentrated feed. One of our mares developed a preference for barley overnight and didn't like the taste of oats any more. Another mare only wanted oats and turned her nose up at barley. We tend to indulge these likes and dislikes, partly in order to meet our mares' nutritional needs but partly because we love them.

Some mares also show colic-like symptoms during this stage. We normally assume that abdominal pain in late pregnancy is likely to be caused by the foal pressing on the mare's intestines, but we always consult the vet in these cases.

Around this time some mares develop a worrying discharge accompanied by blood or mucus, which disappears by itself after a few days. This

This mare is close to foaling; her pregnancy is hard to miss.

should be investigated and monitored by a vet. A mare's drinking behaviour may also change. Many pregnant mares don't like to drink cold water because it triggers movements of the foal. In a mare close to foaling, these movements can be seen clearly just after drinking, or after being moved to pasture.

Mash recipe (for warmbloods)

- *1 kg crushed oats*
- *1 cup linseed, simmered for ½ hr, or*
- *1 cup ready-prepared linseed, soaked in ½ l hot water*
- *1 cup wheat bran*
- *Stir, cover and leave to stand for 1 hr*

Feeding and worming a mare close to foaling

The mare's requirements in terms of energy, protein, calcium and phosphorus begin to rise from the eighth month of pregnancy. These needs can be met by feeding good-quality hay and haylage, together with supplementary feed designed for brood mares. Top-quality feed is essential. Digestion promoters such as mash, molasses and linseed support intestinal activity.

Mares in foal should be wormed regularly – just like the rest of the herd – using products suitable for use in pregnancy. The literature on worming recommends an additional treatment around two days before foaling. Threadworms in particular can be ingested by foals from their mother's milk, or through wall licking or contact with contaminated bedding. However, worming treatments can affect the digestion and general well-being of the heavily pregnant mare. This is why we delay worming until the day of foaling itself, once the mare has recovered from giving birth.

Heavily pregnant mares should not be overfed just before foaling. Feed rations should be divided into several small portions with slightly less forage being fed. Mares should still be receiving an adequate supply of minerals.

Here too, however, some mares may reject a mineral feed that they would otherwise eat happily. There are many different types of mineral feed, so there is bound to be one that the mare will like. Carrots are a healthy, well-tolerated source of carotene and vitamin A. We feed a maximum of one to two kilos per day. Easily digestible mash can be given once every two to three days.

Mares should be fed with restraint immediately after foaling. What they need most at this time is plenty of good-quality forage to make up the volume deficit in the abdomen. We feed mash daily and divide the forage into multiple portions throughout the day. Due to contraction of the uterus, some mares experience colic-like abdominal pain in the first few days after foaling. This can last until their foal heat. A vet should be called promptly to inject antispasmodics or painkillers.

Herbal remedies

Although the use of homeopathy is questioned by the scientific community, the authors propose the following herbal remedies, however, if owners have any medical causes for concern relating to the care of their horses – especially pregnant mares – veterinary advice should always be sought first.

Raspberry leaf tea can be fed in the last two months of pregnancy. Give the mare a teaspoonful of dry tea or a cupful of tepid tea, including the leaves, in her feed.

Juno's belly has already dropped. The birth of the foal is imminent.

A useful herbal remedy to regulate contractions and to ease foaling, or for a difficult birth, severe afterpains or injury to the birth canal, is Caulophyllum C30. Give three drops once a day, starting from around a week before the mare's calculated due date; if necessary, continue for up to three days after foaling.

A remedy which promotes healing of the uterus and is useful for colic-like afterpains, mild post-partum bleeding or a retained placenta, is Sabina C30. Give three drops twice a day after foaling, in addition to any veterinary treatment, until an improvement is seen; in acute cases, give every 15 minutes.

The birth of the foal

Foaling season is the most exciting time of year for any breeder. Every foaling is special and eagerly anticipated.

Signs of the approaching birth

When mare and foal are ready for the birth, this is announced by a variety of external changes in the mare. Her belly drops, the broad pelvic ligaments

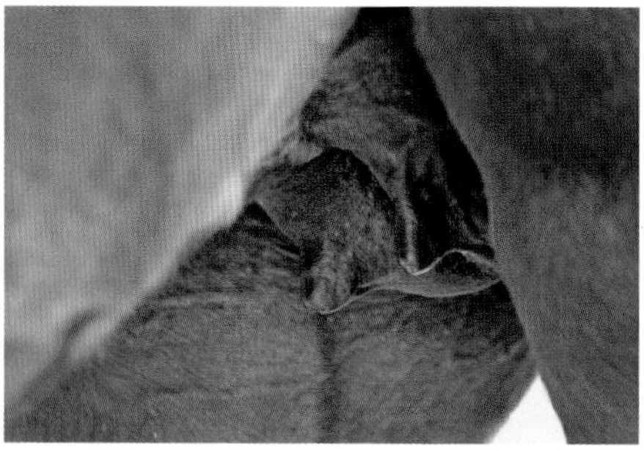

Often, one half of the udder is larger than the other. In this case, the foal has already emptied the left side. In fact, the teat is still glistening.

relax a little, the udder – which has grown steadily in recent weeks – becomes fuller and tighter, the labia are slightly swollen and in some cases the vulva opens slightly when the mare is lying down. The mammary glands start to produce colostrum, which takes the form of small yellowish drops of fluid at the ends of the teats ('waxing'). In the worst-case scenario, colostrum may appear several days or sometimes even weeks before the birth. This causes some of the vital early milk to be lost. If this loss of colostrum goes on for only a few days there will still be enough gamma-globulins (antibodies) left after the birth to protect the foal. But if the dripping or spraying of colostrum continues for longer, the foal will initially need to be bottle-fed. Breeders should always keep a stock of deep-frozen colostrum from another mare on hand for this contingency. Nowadays there is also the option of using artificial powdered colostrum. However, it is better to use colostrum from your own mares, as this will contain antibodies from the mare's home stable.

In some mares, the udder is already very full while the teats remain lax, and no waxing is visible. The milk doesn't come in until the birth is under way.

Owners are often advised to handle the udder of a heavily pregnant mare to get her used to being touched. We do not recommend this. It is all too easy to stimulate the milk flow, causing a loss of valuable colostrum. Before pregnancy or in its early stages, on the other hand, it is quite safe to touch the udder. (For more about colostrum, see pages 13 and 22.)

WHAT TO DO BEFORE THE BIRTH

- Heap the bedding up around the walls of the stall
- Remove the salt lick
- Make up an emergency kit of baby bottles, teats and powdered colostrum and keep it at the ready
- Prepare a foaling kit
- Get a container ready for the afterbirth
- Deposit enough straw outside the stall to provide clean bedding after the birth
- Cut some pieces of string to tie up the afterbirth

Foaling kit

I always make up a foaling kit well in advance of the mare's due date. The kit contains an umbilicus disinfectant, disposable gloves, a single-use enema (get this from your vet), a pocket torch or head torch and two towels.

The foaling kit should be ready well in advance!

LOSS OF COLOSTRUM

One of our mares is prone to losing colostrum. In her first pregnancy, the milk flow started two weeks early and there wasn't enough colostrum for the foal. Lack of antibody protection led to the foal developing a high fever, and it nearly died. Fortunately it survived thanks to veterinary treatment and feeding with artificial colostrum.

The year after that, the mare in the next stall foaled six weeks before this mare. This clearly upset her hormone balance. She started to produce milk and lost a huge amount of colostrum. But luckily the milk flow stopped after a fortnight or so and she still had enough colostrum left when her foal was born.

This mare's mother also lost colostrum before foaling, so this seems to be an inherited trait. In any case, a supply of frozen colostrum should be kept on hand as a reliable source of first aid for the foal.

Keeping a vigil

As foaling draws closer, inexperienced breeders are naturally anxious and keen to get everything right. As flight animals, horses are able to delay birth for many hours. They wait until they feel safe. Noise, clumsy helpers, nervous owners, strange dogs or stormy weather can unsettle them. Especially for skittish mares, peace and quiet are essential for a complication-free delivery.

This is why most births happen late in the evening or at night.

Camera and birth alarm

If complications develop during foaling, you need to be able to step in as quickly as possible – and to do that, you need to monitor the mare. There are some excellent and reasonably priced wireless surveillance cameras on the market nowadays. They allow mares to concentrate on the birth process undisturbed, surrounded by their friends, while their owners watch on the monitor from the warmth of indoors.

We used to take turns to check the monitor every half an hour. Despite this, we sometimes found that the newborn had made its arrival unobserved. In a textbook case the expulsion stage takes up to 30 minutes; clearly, the mare had foaled in between two checks. Now, we fit our mares with a birth alarm belt equipped with a transmitter. This sends an acoustic signal if the mare lies flat on her side, signalling the onset of expulsion contractions. The safest option, therefore, is dual birth monitoring using a camera and a birth alarm. Both of these devices are available to hire!

NORMAL FOALING PARAMETERS	
Dilation stage	50–80 minutes
Expulsion stage	max. 30 minutes
Afterbirth stage	ideally 30 minutes
Length of foaling	1.5–3 hours
Expulsion of afterbirth	within 2 hours
Time for foal to stand	approx. 1 hour, max. 2–2.5 hours
Time for suckling reflex to develop	max. 20 minutes
First drink of colostrum	ideally within the first 2 hours
Passing of meconium	the sooner the better, but within 12 hours
Urination	6–10 hours after birth
Imprinting period	up to 4 days
Heart rate	immediately after birth: 70 beats per minute, 1 hour later: 130 beats per minute, then approx. 90 to 100 beats per minute

A camera and birth alarm belt allow you to monitor the mare from indoors.

Mares in labour become restless and start to sweat. They paw the ground, swish their tails, lie down for a while and then stand up again. Contractions are separated by quiet intervals in which mares may even start to eat again. Mares differ in the intensity of their response to contractions. As long as the waters haven't broken, the intervals between contractions are not harmful. They give the mare a chance to relax and find a good spot to deliver.

We stay out of the stall until the foal has begun to emerge through the pelvis, and even then we only enter to check that the amniotic sac is torn over its nostrils. Otherwise, we leave the mare to it.

An ideal, complication-free birth takes around one and a half to three hours in total. In the dilation stage, the cervix opens and expands. The water bag widens the whole birth canal.

Breaking of waters, expulsion stage

When observing a birth, the breaking of waters is impossible to miss. Noisy and copious, it marks the transition from the dilation stage to the intense expulsion stage. Mares normally lie flat on their sides, pushing and groaning, straining and stretching out their legs. But they also stand up again from time to time. Lying on their sides creates the best conditions for mechanical expulsion. After all, the heavy foal, which weighs up to a fifteenth of the mare's weight, has to be shifted towards the pelvic opening from its position deep inside the mare's belly.

Before birth, the first uterine contractions turned the foal into the ideal position for delivery. Its head is wedged between its forelegs. The underside of its hoof points towards the mare's udder. The edges of the hooves are naturally covered with a soft down, a protective mechanism that prevents injury to the birth canal. Often, one foot emerges before the other. As a general rule, mares don't need any outside assistance. Never try to help by pulling

COMPLICATIONS

Despite our spacious foaling stalls, we have still had a case where an experienced brood mare lay down so close to the wall that the foal did not have enough space to come out. During a painful labour, the mare had burrowed deeply into the straw and become stuck. The foal's legs were already emerging.

With several helpers, we managed to level out the straw as the birth progressed and carefully manoeuvred the mare more into the middle of the stall. In the end, the mare stood up and found herself a better spot for the birth of her foal. Especially in a precarious situation like this, it is important for everyone involved to suppress their anxiety, to talk to the mare reassuringly, and not to make any sudden moves but to radiate peace and calm. Our mares know and trust us and I believe they understand that we are trying to help them.

Precious moments of first contact.

on the foal's legs. Only assist if the mare is having a difficult birth and has lost the strength to push. Make sure to synchronise your assistance with the contractions. Pull only during a contraction and only towards the udder, not towards the tail. Ideally, use a towel to improve your grip on the foal's slippery legs.

However, birth complications in horses are rare. If they do occur, they are usually life-threatening for mare and foal. Call your vet as soon as possible if the normal foaling parameters on p. 15 are not in order.

Possible complications

- *Malpresentation*
- *Birth interrupted by external disturbances*
- *Narrow birth canal*
- *First-time mother, large foal*
- *Weakened mare*

After the birth

Once the foal's shoulders (the widest part of its body) have cleared the pelvis, most of the work is already done. The amniotic sac is extremely compact. If it doesn't tear by itself, you need to step in! It needs to be broken and cleared from the foal's nostrils so that it doesn't suffocate.

Mares usually stay down for a while after giving birth, especially if it is not their first time. In their excitement about the newborn foal, maiden mares often don't realise just how exhausted they are.

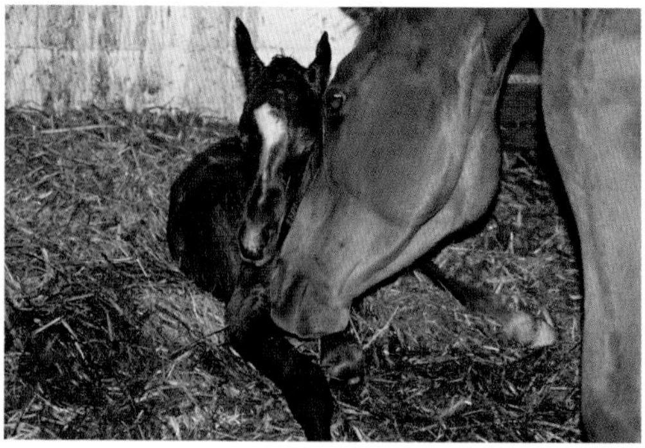

Just born Starlight is still very wet.

The foal is still connected to its mother, and being nourished, via the umbilical cord. Never cut the cord, because the foal will receive another one to two litres of maternal blood before the mare stands up and the cord tears by itself. So if your mare decides to stay down for a while longer, there is no cause for concern. On the contrary, as owners and breeders, these are moments in which we can simply take a deep breath and observe. Our calmness transmits to the mare and her newborn foal. This is the most important thing we can give them at this time.

WHAT TO DO AFTER THE BIRTH?

- Tear open the membrane sac and clear the foal's nostrils
- Treat the umbilical stump with disinfectant
- Colt foals: administer an enema (see p. 25)

WHAT CAN YOU SEE?

- Can the foal stand up?
- Does it pass meconium quickly? (see p. 24)
- Is the suckling reflex present?
- Is the foal drinking?
- Is the mare's vagina torn? If so, inform your vet

WHAT TO DO IN THE AFTERBIRTH STAGE?

- Tie up the afterbirth
- Put the afterbirth into a container and check that it is intact
- If in doubt, ask your vet to check the afterbirth
- Is the foal imprinting normally?

The first few hours

Healthy, vigorous foals often whinny soon after birth. The dam answers her foal's call with a distinctive slow, gentle nickering. The imprinting process between the pair is underway. All other mares in the stable or out on pasture will also answer and greet the new foal. This is always a very moving process.

When the mare stands, the umbilical cord breaks at a certain point around 3 to 5 cm from the foal's abdomen. We treat the umbilical stump immediately using a disinfectant solution, repeating this treatment several times.

The mare normally engages with her new foal in a very caring and motherly way. She nuzzles it and licks it all over. At my first foaling, I actually worried that the mare might nibble off one of the foal's ears! Even first-time mothers have these maternal feelings. The mother's body secretes a hormone called oxytocin, which has effects on milk production, milk let-down into the udder, delivery of the placenta and stress and pain reduction. The release of oxytocin is promoted by pleasant, relaxed contact.

Intensive licking by the mare stimulates the foal's circulation and is a vital part of imprinting. The foal's impressively long legs are nicely on show here.

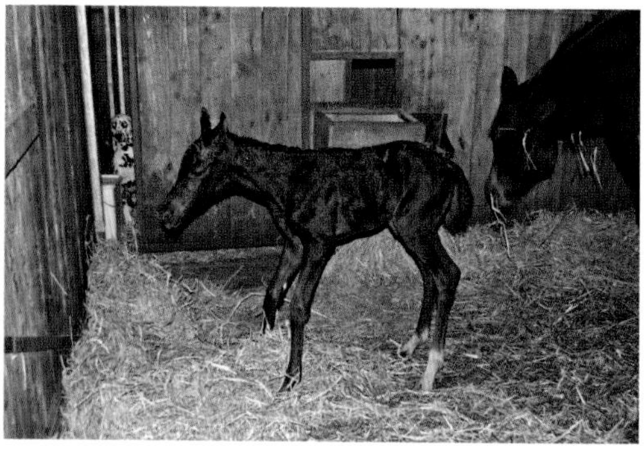

The foal's first attempts to stand are very wobbly. Its brain stores the sequence of movements and activates its sense of balance.

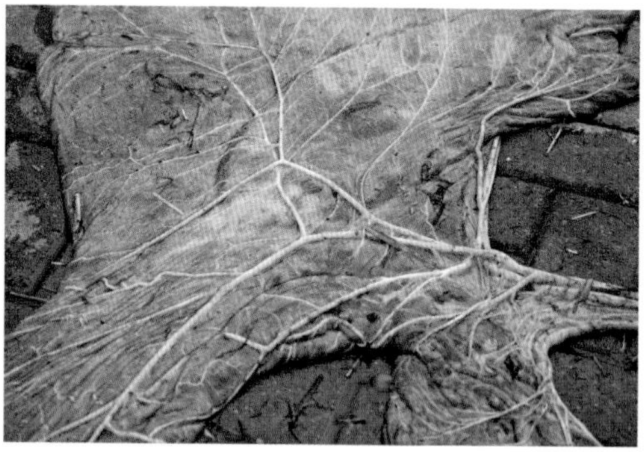

a) Afterbirth b) Foal's bread

Finding the udder

In a healthy foal, the suckling reflex is immediately present and recognisable. Even before they stand, healthy foals will curl up their lips and reach out into the air or against any object that touches their muzzle. This milestone can therefore be ticked off as a sign of a healthy delivery.

As soon as foals are able to stand and walk a little, they start to look for food. Getting the hang of walking is easiest out on open pasture, but even indoors a foal will wobble purposefully towards its mother on still uncoordinated legs and attempt to nudge and suckle her. Foals are genetically conditioned to look for a place that replicates the feeling of standing closely under their mother's body and a particular angle formed by the udder and hind legs. Unfortunately, sometimes they also find this imagined treasure trove between the mare's front legs. Mares, especially the more experienced ones, will present themselves invitingly to the foal time and time again with enormous, unswerving patience, sometimes even lifting up a leg. If the foal happens to brush against the teats, even the slightest touch is enough to make milk spray out from the bulging udder. Now the foal develops a taste for it and is reluctant to leave.

However, sometimes the foal needs to rebalance itself, or it falls over or accidentally wobbles off on its long legs. Then the treasure hunt starts all over again . . . Once the foal is in the right place, the mare rubs it gently on the croup. Sometimes she even raises her belly invitingly, all the time talking to the foal in a special soft tone of voice.

First-time mothers are not always so patient. They are sometimes nervous and irritable, especially if the foal is still a bit clumsy. They keep walking off, then the foal walks off as well. Unsure of the situation, it starts to search somewhere else along the walls of the stall, and a nerve-wracking cycle begins.

The birth was evidently strenuous for the mare as well.

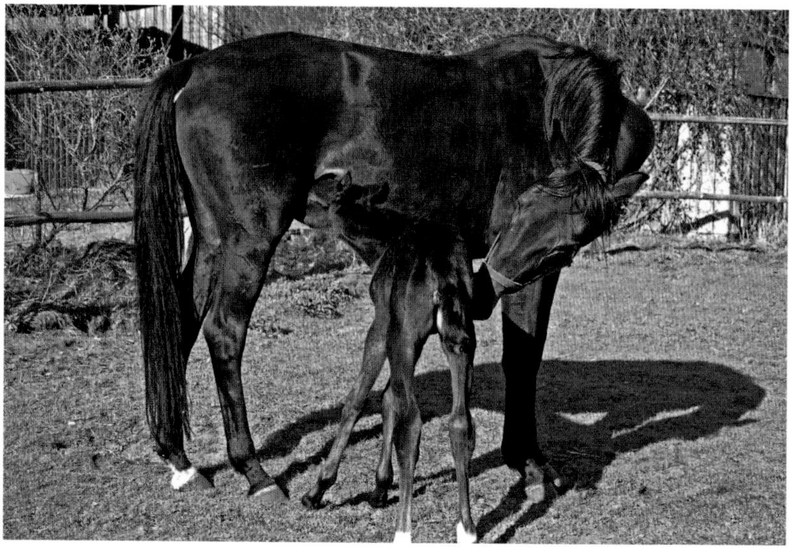

Only a few hours after the birth, mother and foal are together in the paddock for the first time. The mare investigates her newborn foal with interest. These are special moments of intimacy.

But keep your cool and radiate calm. Your little protégé will learn and remember his way to the udder only by finding it himself. Avoid disturbing the imprinting stage and the 'getting to know you' process between mother and foal, for example by guiding the foal to the right place. Any attempt to assist healthy animals delays the first intake of nutrition! As difficult as it might be, stay outside the stall and observe mother and foal from the aisle or by watching on CCTV.

The whole 'finding food' stage can take up to three hours.

Colostrum

The first drops of the mother's milk, drunk in the foal's first three hours of life, are vitally important. They act like a protective shield. A foal can't survive without them. The power-packed first milk or 'colostrum' contains antibodies called 'gammaglobulins' which combat the various disease-causing factors that the mare has confronted and overcome in the past.

However, these immunising molecules are relatively large and can pass through the lining of the foal's intestine only in the first 12 to 24 hours. The mare's milk contains extremely high levels of antibodies for the first seven days. Later on, it contains specific antibodies, is ideally tailored to the foal's life stage and living conditions and changes composition depending on the foal's needs. Colostrum also promotes the passing of meconium (see p. 24).

NURSING HURDLES

One of our mares, a first-time mother, was confused and extremely twitchy after the birth. Each time her filly foal tried to suckle, the mare squealed and dodged aside. Once the foal was at her head end, she calmed down again. The foal had been born very early, after ten and a half months of pregnancy. For three long hours the filly persevered with her quest, displaying huge vitality, strength of character and an iron will. Small and delicate as she was, she did not lie down even once during the whole time. In such cases it is easy to lose patience with the mare. Even 15 minutes of watching a process like this is enough to make you strongly tempted to intervene.

After consulting our vet, we gave the mare a mild sedative and the foal nursed successfully at the first attempt. To build up the sturdy, tenacious

Even today, Early does not tolerate pressure. She expects generosity and polite manners from her rider and gives 100 per cent in return.

little fighter and provide her with her first natural immunisation, we bottle-fed her with some colostrum milked from a mare who had foaled a few hours previously. After nursing again, she was able to lie down and rest.

This filly foal is now a four and a half-year-old mare and her strength of will and fighting spirit are legendary.

Weak foals

If the newborn is too weak or too ill to drink unaided, colostrum should be milked off and administered to the foal, ideally hourly but at least once every two hours. Bottle feeding can continue for days or weeks. If you have never done it before, it is best to seek assistance. Incorrect feeding technique can lead to milk getting into the lungs, which may trigger pneumonia.

FINDING THE UDDER

The last colt foal born to one of our foundation mares was very weak. Try as he might, he simply couldn't stand up. Once every hour, we lifted him to his feet and took him to the udder. The experienced mare just stood there and let him drink in peace. Three days later, I went to the stable to help with another mealtime and the little colt was already standing. I asked around to find out who had helped him, but this time he had done it all by himself. We were no longer required!

Meconium

In the first few hours after birth, a healthy foal passes a substance known as meconium. This is waste material that collected in its gut over the course of the pregnancy. Meconium takes the form of a black stool, up to a metre long in the horse, which is divided into small hard pieces.

Watch out for meconium retention. This is a particular risk in weak foals, colts, foals born after longer pregnancies and foals that were slow to drink their first colostrum.

Colts have a narrower pelvis than fillies, which makes it easier for meconium to build up. This is why we give our colts a ready-to-use enema as a preventive measure. We also treat any fillies that we see straining to defecate.

Take care!

Meconium retention can prove fatal for foals. So it is vital to make sure that the foal passes all of its meconium within the first few hours of life. Repeated writhing or straining is a sign of retained meconium blocking the gut.

Signs of meconium retention in foals include colic-like pain, rolling and kicking at the abdomen. The pain on their faces tells you they are feeling uncomfortable. Abdominal pain demands a foal's full attention. The sooner it passes the meconium, the sooner it will be able to focus on finding the teat, learning to walk, imprinting on its mother and dealing with its new world.

Meconium retention needs to be differentiated from a physical defect of the urachus duct that stops urine draining normally via the urinary tract. The result is that urine collects in the abdominal cavity. The foal strains in the same way as it does when defecating, and displays colic-like symptoms similar to those of meconium retention. So it is important to check that the foal passes both meconium and urine during the first few hours of life. Foals should not urinate from the umbilicus!

Meconium looks like a string of shiny black pellets. Here, it is already coated with the yellow milk faeces.

Administering an enema

The enema is warmed to body temperature in a container of hot water. You need a practised assistant to help administer an enema carefully and avoid the risk of perforating the wall of the foal's gut. We work in pairs when giving an enema, usually with the foal lying down. I wear a head torch for better visibility when gently inserting the warmed enema. Inexperienced breeders should always seek expert assistance!

When restraining the foal, take care to avoid bending its tail up as this is painful and risks causing spinal cord damage.

The afterbirth

The mare should pass the afterbirth or 'placenta' within two hours after foaling. Once the mare gets to her feet, the amniotic membranes hang down between her legs. Tie them up using the pieces of string you cut earlier. This will stop the mare from standing on them or kicking at them in irritation. The mare often lies down during this stage; she is experiencing afterbirth contractions as the placenta detaches from the uterus. Right at the end of the process, the afterbirth's own weight helps the mare to expel it completely. Take care never to pull on the afterbirth!

Delivery of the afterbirth can take a while. Here, the afterbirth has been tied up to stop the mare from standing on and tearing it.

Place the afterbirth in a lidded container that you got ready earlier. This is so you can check that the placenta is intact. If any parts have been retained in the uterus, your vet will need to remove them. Consequences of a retained placenta include fever, inflammation (possibly making the mare unfit for breeding), a drastic drop in milk protection and post-foaling laminitis. These in turn can lead to a sick mare who is unable to take care of her foal.

Some mares experience strong afterbirth contractions. These colic-like pains can be severe enough to prevent them from taking care of their foals. If this happens to your mare, ask your vet to administer an antispasmodic.

Cleaning out the foaling stall

Once the mare and foal are healthy, happy and well cared-for, remove the wet bedding and top up with fresh material. As you clean out the stall, you may find a liver-like substance called the hippomane or 'foal's bread'. This is an accumulation of metabolic residues which floated around in the amniotic fluid during the pregnancy. It tends to break down and disappear rapidly after birth. Finding the foal's bread depends on close monitoring of the afterbirth stage. In the past, stable masters were sometimes rewarded for finding the foal's bread, as it proved they were keeping a careful watch on their mares.

When preparing the foaling stall and mucking out after the birth, make sure to heap the bedding up around the walls to stop the foal bumping into or falling down against the bare wall when it makes its first attempts to stand up. The foal is still very front-heavy so it keeps pitching into the straw head first.

Once you have cleaned out the stall, leave the mare and foal alone. This allows them to interact and gives the foal a chance to work on its standing, walking and drinking.

The imprinting stage

Newborn foals take a while to become clearly imprinted on their mothers. As animal behaviouralist, Konrad Lorenz, demonstrated in goose chicks, animals do not have an inborn idea of their own species.

In the first few hours, foals will follow a person who spends too long in the stall. But once they have nursed, they often become timid. After their first night alone with their mother, most foals have clearly imprinted on her and know her voice and smell.

The foal has never seen fences before, so we lead the mare and foal on the first day without letting them go free. We always make sure there are at

Galaxy and Universe were taken outside within hours of the birth. Gentle exercise is good for both mare and foal.

A healthy foal is bold and lively.

least two handlers. After the close confines of its mother's belly, the foal is amazed by the vastness of the outside world. The mare may like to nibble some grass or have a roll. Depending on its temperament, the foal may try out its paces but it always sticks close to its mother. The mare can be trotted in hand with the foal cantering alongside. We make sure that the foal is not taken too far away from its mother, or the mare will become agitated. All other mares and horses in the stable participate in this first excursion, like an audience watching a show. Every gaze is in the same direction as they look on nervously and whinny to greet the new arrival. The mare is totally preoccupied with her foal.

Relaxed togetherness

In the days that follow, we leave the mother and newborn in a grass paddock of their own for a few hours, within sight of the mare herd. This gives them the nearness and protection of the herd without the stress that can be caused by other inquisitive mares or young horses. They have exercise and fresh air, and the foal can get used to being fenced in. It takes around four days for the foal to imprint fully on its mother. After this time, the pair are introduced to the mare herd. Later on, mother and foal sometimes lose sight of each other. The foal naps or plays and forgets to keep track of its mother. Then it whinnies for the mare and, if she doesn't reply, starts to look

Growing is hard work, sometimes you need a break!

for her. If the mother is a chestnut, the foal approaches the nearest chestnut mare and tries to make contact through smell. The mare signals clearly 'stop, I am not your mother', and the foal seeks out further chestnut mares until it finds its mother. Searches like these are usually calm and stress-free.

Foals have different degrees of tolerance to touch. Some will let you touch them anywhere from the start, while others will only allow their mothers to touch them after the imprinting stage. Tolerance to touch is a character trait that the foal retains after people have built up contact with it. Its influence is still apparent years later.

TIP

Every day offers constant opportunities for a foal to learn what the modern horse needs to know. Lessons continue for 24 hours a day.

Imprint training

On many stud farms, especially in the USA, it is common practice to restrain foals in the first few hours after birth and to touch them systematically all over, including in orifices such as the mouth, ears and anus. If the foal protests, the touch is repeated until the foal stands still. Foals are also exposed to strong visual and aural stimuli. This shows an appalling lack of respect for a newborn creature and its mother, and we are strongly against it!

Even very young foals will try a nibble of grass.

Foals are adventurous and don't stop at the pasture gate simply because their owner wants them to. This calls for foresight!

Recent research confirms our 'gut feeling'. For example, the method known as 'imprint training' (Miller, 1992) can trigger learned helplessness through its use of sensory over-stimulation (Zeitler-Feicht, 2001). Studies comparing 'treated' and 'untreated' foals over extended periods have shown that treated foals were more anxious and shy with people, and also more insecure and less playful with other horses (University of Rennes, according to Kavallo magazine, Switzerland 2011).

Management and feeding of the nursing foal

Foals should grow up as naturally as possible. Right from the start, they need plenty of fresh air and exercise on firm, dry ground. Given these conditions, minor conformation defects usually self-correct in the first few weeks. Even winter-born foals should have at least an hour or two of outdoor exercise with their mothers.

All foals have a natural urge to move that they should be able to act out. But exercise is important for the mare as well. If she keeps moving, she will 'cleanse' much better, her uterus will contract faster and the draining of wound secretions will be improved.

A foal with weak flexor tendons. Rolli's tendons tightened up after a week or so of unrestricted exercise on firm ground.

In contrast, a foal with contracted tendons ('knuckling over'). After three days in plaster bandages, Memory's fetlocks were in the right position. However, the plasters didn't stop him cantering around.

Feeding to the foal's needs

Vital colostrum is the first food that a nursing foal receives (see also p. 22). Foals nurse around 50 to 60 times a day, and our Trakehner foals drink 150 to 250 millilitres of milk at each feed. Even at only a few days old, they are already nibbling on hay or haylage, or trying some grass or their mother's concentrates.

The wide-legged grazing stance takes a bit of practise . . .

After the first few weeks, mother's milk alone is no longer enough to sustain them. If foals are introduced to lush spring pasture during this time, their first solid food will be fresh meadow grass. Winter-born foals, however, are fed a foal supplement in a special creep manger designed to stop the mares eating their ration.

In the spring, once the growing season has started, the mares can be given at least an hour or two on pasture with their foals. The time on spring grass is increased gradually to get their digestive systems used to the change in food source. Spring grass is very rich in protein and low in crude fibre.

Summer pastures no longer supply enough minerals and calcium for nursing foals to grow healthily, so they are also given an appropriate supplementary feed.

Note: read about overfeeding on p. 89.

Be like a good brood mare

Protein-rich spring grass is an ideal supplement to mother's milk

The idea of the dominant alpha human, the boss at all costs, is unfortunately still all too prevalent on the equestrian scene. Constant bossing around of subordinates and an insistence on preserving hierarchies are characteristic traits of animals that fear a loss of authority. Real authority is not open to question! Your impressive, solid presence and your sense of responsibility for the well-being of the whole herd leave no doubt as to your trustworthiness.

The impression that lead animals are particularly dominant came about not through observing actual wild horses but under unnatural conditions characterised by outdated approaches drawn from the field of behavioural biology. Even semi-feral Konik stallions, who like plenty of space, join forces in winter in order to save resources.

Foals are raised by the whole herd, but mostly by their mothers. Mothers are their foals' lead mares. To raise your own foal successfully, watch closely how its mother and the rest of the herd behave. How does the rest of the herd behave, and who does your foal or young horse trust?

A GOOD BROOD MARE . . .

. . . is always there for her foal
. . . is patient with her foal in the early days
. . . does not need to keep demonstrating her status
. . . sets clear limits
. . . offers comfort
. . . provides food
. . . provides protection
. . . is reliable
. . . is predictable
. . . is fair
. . . is alert
. . . is clear and consistent
. . . expects the clear messages she gives her foal to be followed promptly
. . . is not a substitute playmate

Notice how unaffected this mare is by her foal's escapades as he shows off his talents as a dancer. Without feeling provoked or wondering if she is still the boss, Fidelity allows her wild child to be himself.

The foal's character

Our brood mares differ widely in how liberal they are in raising their foals. This has nothing to do with the mare's rank. Some mares won't allow their new foals to leave their side in the early days and weeks. They either call the foal back immediately or keep chasing after it. Many of these over-protective mothers are slightly older mares which took part in sport for many years and were not used as brood mares until later in life.

Other mares allow their foals to go off on journeys of discovery after only a week or so. They watch their foals closely of course but, provided they can see them, they leave them to it. Later on, these foals are bolder and more open to new things than foals that were closely guarded. Over the initial years of rearing and training, the differences between the two types become less marked. However, the foal's original character keeps resurfacing in critical situations. In a herd of young horses, the ones that were raised by their mothers to investigate their surroundings calmly are always the boldest, while the foals of the over-protective mothers tend to stand out on account of their cautious natures, irrespective of their rank in the herd.

What foals should learn

Foals are always learning, and not just in their official lessons! Their main teachers are their mothers. A trusting relationship between foal and mare is the best basis for the image that your foal will have of humans.

Young foals have a much greater capacity for learning than older foals. So use this early time to practise basic exercises, without stressing the foal. Repeat the exercises occasionally, so that they become ingrained and automatic. You will be surprised at how quickly your foal gets a grasp of new ideas.

Practise little and often, and end on a high. Exercises should make your foal want to be with you and learn with you. The work is over and done with before it really gets started. Like children, foals have only a short concentration span but they learn quickly and easily. Two positive minutes are better than ten average ones.

As you do each exercise, check your own mood, breathing and posture. Are you truly relaxed? Are you breathing calmly and evenly? Are you standing comfortably, shoulders relaxed? Horses are masters at decoding body language. That's how they have managed to survive this far! If we are relaxed, that transmits to the horse.

If your foal was calm at first but then becomes unsettled, stop what you are doing. Tickle your foal in a favourite spot, or leave it in peace for a while and try again later.

Remember that an exercise that works on pasture will have to be practised from scratch in the paddock. This also applies if new or different people are present.

Only try an exercise if you yourself are confident with it. Plan how you are going to approach the exercise and create ideal conditions for learning.

In principle, our view is that the ideal way for foals to learn normal horse behaviour is in the herd together with their mothers. The grazing season, from spring through to autumn, is the best time for this.

Learning is easiest in the herd, where it takes place informally and in small chunks. But remember your own safety and keep an eye on the herd dynamic by looking around in every direction.

EXERCISE 1: MAKING CONTACT AND BUILDING TRUST

Aim of the exercise

A foal starts to form its picture of the world from day one. As humans, we need to integrate ourselves into that picture in a positive way.

What you need to know

A foal builds up trust indirectly via its mother. It watches how you handle its mother and senses the mood, the relationship, between you. Hopefully, it learns that you are the nicest person in the world and there is nothing to fear from you. Direct contact through touch is important. At this stage foals are still small and easily handled, but even a one- or two-month-old foal is much stronger than a human being.

Stroking with the whip – a trust-building exercise that is also invaluable in adulthood.

How to approach the exercise

Start building trust and making direct contact with the foal on the day after its birth. Even shy foals are inquisitive – so much so that they will approach and start to investigate you if you crouch down calmly in their stall. Be aware of your own safety but stay loose and keep your breathing relaxed. Nearly every foal will let you touch it after only a short time.

If it keeps its distance, spoil its mother with some carrots or apples or by showing her affection. Foals learn from their mothers' behaviour, so this should finally convince it to trust you.

If your foal is shy

If a foal is very timid, you could try some gentle stroking with the whip. Once again, start with the mother before you include the foal. The advantage of stroking with the whip is that you can maintain some distance between yourself and the foal and the touch feels less intense than touching with your hand.

Establishing this initial contact can take a long time in some cases. For a more detailed description of this exercise, see page 77.

EXERCISE 2: LIMITING FREEDOM OF MOVEMENT

Aim of the exercise

To give your foal a brief sensation of having its movement restricted.

What you need to know

Initially, young horses take unlimited freedom of movement for granted. Newborn foals do get to know the walls of their stall but often they don't recognise fences for the first few days, unless the fences are really sturdily built. Allowing someone to restrict its freedom of movement is an important stage in a foal's learning.

Precondition

Only restrain a young foal once it trusts you enough to let you touch it all over. This can take a while for shy foals. The mare should respond calmly and trustingly when you approach. If the mare gets worked up, the foal won't stand still either!

Visit and stroke your foal regularly!

How to approach the exercise

You can attempt this exercise alone but a helper can be very useful. Approach the foal from the front and side, and tickle it in a favourite spot. Stand beside the foal and put your arms around it, with one arm in front and the other behind. As you do so, tickle it in places where you haven't tickled it before. Restrict the foal's freedom of movement using the arm in front. With your other hand, touch the area to the side of the tail root or underneath the tail. If there are two of you, one person can hold the foal from the front and tickle it while the other holds and tickles at the other end. Take care never to bend up the tail! The foal enjoys your touch but will eventually feel an urge to escape. Hold it calmly for a moment longer before you let it go.

EXERCISE 3: WEARING A HEAD-COLLAR

Aim of the exercise

Wearing a head-collar is a basic exercise.

What you need to know

We describe this exercise in detail specifically for shy or timid foals. Laid-back foals will accept a head-collar easily without the rope preparation stage.

Pandsy accepts the head-collar first time without any problems. Rhapsody watches and learns.

Be careful around the touch-sensitive poll area.

How to approach the exercise

To accept a head-collar, your foal needs to be trusting enough to allow itself to be touched all over, including on the head, face and ears, and you need to be able to restrict its freedom of movement briefly. Start by simply placing a rope loosely around the foal's neck. Slide the rope up as far as its ears. In a second stage, lay the rope across the foal's nose and knot it under its chin. Do this quickly and playfully, making sure that the rope can be quickly taken off again. After repeating this exercise a few times, you can try it with a real head-collar.

EXERCISE 4: GIVING A HOOF

Aim of the exercise

To teach your foal to feel safe and secure on three legs and to familiarise it with tools such as hoof picks and rasps.

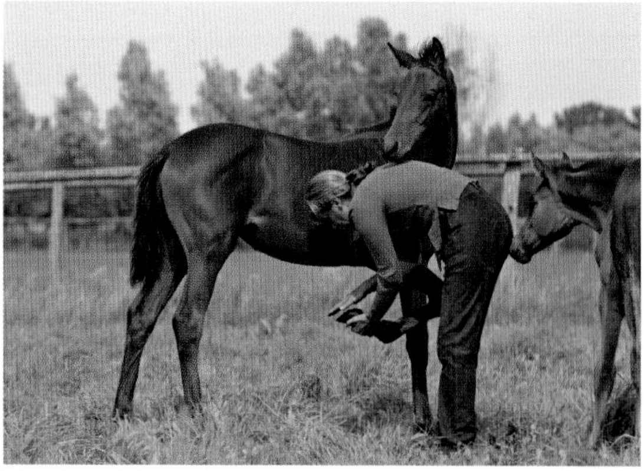

Pandsy is already basically trusting of people and open to new experiences. Even tapping her hoof with a hand doesn't frighten her.

What you need to know

Foals have not yet developed a good sense of balance. They find it hard to stand on three legs. They may be afraid of falling over so refuse to pick up a hoof. There seems to be a sort of reflex point on the foal's hock that, when touched, immediately causes the animal to lash out. This reflex subsides after it has been triggered a few times. Never hold on desperately to the foal's leg. The foal's immature bones and muscles are prone to injury. For example, we once had a foal that ended up with a blocked sacroiliac joint because a farrier wouldn't let go of a flailing leg.

Precondition

Your foal must be trusting and allow you to restrict its movement, touch it and stroke it with the whip (p. 77).

How to approach the exercise

To get your foal used to having its legs and feet touched, stroke its legs with a whip, or even two whips. If it accepts this, progress to stroking its legs with your hand. Some foals accept a touch with the back of the hand better than a touch with the palm. After stroking briefly with your hand, lift up each leg in turn, working quickly and calmly. The aim is to help your foal to understand that it can balance itself. Sometimes your foal needs to reposition its legs slightly before it will let you pick up a hoof. If this is the case, wait patiently until the animal is standing squarely. Lift the hoof for only a few seconds before gently placing it down again.

If this goes well, the next time you can tap on the hoof sole to mimic the farrier. Another time, you can introduce your foal to the hoof pick or the rasp.

EXERCISE 5: LEADING

Aim of the exercise

To teach your foal, in a series of small, anxiety-free steps, that there are some situations in which it can't react instinctively. It learns the beginnings of self-control.

What you need to know

Leading a foal is a basic exercise. However, a number of things can go wrong. A charging foal that rears or takes off cannot be brought under control if it has not yet learned to walk on a lead rope. A fall can cause serious injury to the foal's spine. Short leading exercises are essential if your summer pasture is not right beside the stable or, for example, if you need to travel to another yard for a foal assessment. Otherwise, more advanced leading exercises – in the stable aisle or outdoors – can confidently be put off until after weaning.

One of Flightshow's first leading exercises. Take care: the knot in the neck rope must not tighten when the rope is pulled.

How to approach the exercise

Once again, practise in the stall first. Besides using a head-collar and lead rope, we recommend that you drape another rope around the foal's neck. You can use this to bring a charging foal safely to a halt and to prevent it from rearing without twisting its cervical vertebrae. While working in the stall, it is enough to give the foal a feel for gentle leading signals and a sense of being restrained. Practise for only a few minutes at a time. Tug gently on the rope to encourage your foal to circle round to the left or right, or to halt. Lead from both sides. Later on, take the mare and foal for a walk across the yard to the pasture or paddock.

EXERCISE 6: PREPARING FOR TYING UP LATER ON

Aim of the exercise

Allowing itself to be tied up is a basic exercise, although it doesn't need to be mastered until the horse is three years old. In preparation for later perfection, you can teach your foal to control his impulses briefly.

What you need to know

We never tie our young foals. Even tying them to their mothers, for example to a girth or even the tail, is a practice we see as an outdated and dangerous relic from the past. The risk of injury is huge if a panicked foal gets tangled up in the rope. The foal's immature bones and ligaments are very vulnerable, and not just in the neck area!

TAKING CARE WHEN LEADING AND TYING UP

From an osteopath's point of view, the risk of injury associated with leading, securing and tying up foals and young horses is significant, as are the consequences.

The whole spine can become blocked, from poll to haunches, but equally so can other joints such as the fetlock, knee and hock.

Besides blocked joints, there are also the soft tissues (muscles, ligaments, tendons) which can become twisted or over-stressed. Nor are the internal organs spared, because they are suspended from the whole scaffold. Spinal blockages lead to reduced blood supply and soft tissue blockages to imbalances in the organs themselves.

If the foal or horse continues to move despite these blockages, which are not obvious to a lay observer, changes can occur in its pattern of movement and gait. The farrier is usually the first to notice these. We can illustrate this with an example in humans: if your pelvis is misaligned, the pattern of wear on your shoes will be uneven.

Gina Seydel, physiotherapist and equine osteopath (DIPO)

Marietta has passed the lead rope though the head-collar in such a way that she can hold both ends of the rope in her hand and could release Flightshow at any time. Using exemplary, non-aggressive body language, she invites four-week-old Flightshow to walk with her.

What you need

Foal head-collar and rope.

How to approach the exercise

Your foal should be trusting and familiar with the head-collar, and should allow itself to be touched. Pass the rope through the bottom ring on the head-collar and hold both ends of the rope in one hand. Now hold the foal still for a few moments. It should notice the restraint but not be afraid. If the foal is unsettled, you can allow him more rope. If he panics, you can release him immediately. As a rule, you only need to 'give' slightly to calm the panic that some foals feel on being restrained. Use your body language to encourage the foal to lower its head. This teaches foals that there is nothing to fear in being restrained.

The Tellington Method

These exercises are based on the Tellington Method, developed over 30 years ago by Linda Tellington-Jones as a unique combination of body work, ground work and riding. The basic principle and aim of the Tellington Method is to create an understanding, trusting, anxiety-free bond between handler and horse. It increases the animal's capacity to learn and emphasises safe, step-by-step working.

EXERCISE 7: MOUTH WORK AND PREPARING FOR WORMING

Aim of the exercise

Foals are wormed for the first time at four to six weeks of age (see p. 49). Get your foal used to being touched inside his mouth before trying to worm him!

What you need to know

For many foals it is entirely natural to touch things, including people, with their mouths. Like all small mammals, including children, they start to explore the world by using their mouths.

Five fingers

The cardinal rule for mouth work is that every finger that goes into the horse's mouth must come out again!

How to approach the exercise

Stand beside the foal's head and look in the same direction as the foal is looking. Hold the foal gently with one arm across its neck. Start by tickling it in one of its favourite spots, for example on the chest or withers. Now

Starlight has been touched inside her mouth before. Now she investigates the wormer with her lips. Soon after this, she downed her dose quite happily.

carefully trace two relaxed fingers around the foal's mouth. Slide one finger inside its mouth, staying close to the gum. End the exercise when you feel the foal's patience start to falter. Practise in short bursts for one to two minutes, repeating the exercise two or three times a week. The next time, stand on the other side of the foal. In all of these exercises, make sure to avoid favouring one side over the over. If your foal is already familiar with a head-collar, you can put one on before doing the exercise.

EXERCISE 8: EAR TICKLING

Aim of the exercise

Stroking a horse's ear relaxes its poll area. Later on, this will help you to lower the horse's head.

How to approach the exercise

Gently massage or tickle the poll area around your foal's ears. Circle with your fingertips in this sensitive area. Apply only a light pressure. Touch the foal's ears as well, and stroke gently towards the tips. This has a particularly relaxing effect on the poll.

EXERCISE 9: TICKLING THE TAIL AREA

Aim of the exercise

Nearly every foal loves to be tickled in the tail area.

Most foals enjoy this exercise. Touching and massaging the tail area is part of typical 'horse talk'. Tail ticklers are friends! Massaging the tail area also prepares your foal for having its temperature taken later on, and for being washed and having cream applied after attacks of diarrhoea.

How to approach the exercise

Place one arm around your foal and with your other hand tickle around his tail. Tickle all the way round the tail itself, including on the underside. If your foal is relaxed, you can also touch underneath the tail around the anus. Just above the anus is an acupuncture point that helps the horse to expel wind when suffering from digestive problems. Press or massage this point several times.

EXERCISE 10: SIMPLE CIRCULAR TTOUCHES

Starlight pricks her ears attentively at these strange touches. Ear massage helps the body to deal with circulatory and digestive problems, itching, muscle soreness, tiredness and many other conditions. Hold the foal's ear between your thumb and index finger and massage gently using small circular movements.

Aim of the exercise

The effect of gentle circular movements based on the Tellington Method (TTouches) is legendary. They promote attentiveness, learning ability, body awareness, self-awareness and joy of movement. They also reduce tension and stress.

How to approach the exercise

Trace small circles all over the foal's body. They can be either interconnected or dotted at random.

Place your hand on the animal's skin and push the skin around for one and a quarter circles. Don't simply glide your hand over the coat, but actually move the skin. This works best if you use only a gentle pressure.

Try it on your own forearm. Imagine you are placing the tips of your index and middle finger on a clock face at six o'clock. Now, rotating clockwise, push the skin under your

The Tellington TTouch is not massage. The touches are feather-light. They are targeted at the nervous system and improve body awareness, mobility and learning ability. They are good for both giver and recipient and strengthen the bond between you and your horse.

Fillies do sometimes rear during play but they are generally more 'grounded' than colts.

fingertips around the face of the clock until you come back to six o'clock, and then keep going to nine o'clock. This is your one and a quarter circle. Good places to start these circles are on the foal's neck or its hindquarters on either side of the tail.

Differences between colts and fillies

The difference in play behaviour between colts and fillies becomes apparent within the first few weeks of life. All foals play running and chasing games.

However, colts seek the company of other colts, preferring them for the play fights and chases that are typical of young stallions. They nip, shove, rear, bite each other on the chest, flanks and forelegs, get down on their knees and practise playfully for later confrontations. They show a typical play face rather than an aggressive fighting face.

Fillies will also play with males and do sometimes rear for fun. But their favourite pastime is a spot of light grazing and intensive coat care.

To grow up happy and healthy, colt foals need the presence of other males. In the foreground is Rolli, whom we met on p. 31 as a foal with lax tendons.

It is difficult for foals if they have to grow up without other horses of their own age. Rearing foals in this way is distressing for both mare and foal and often leads to behavioural impairments.

All colt foals and some fillies nip or pinch each other. If targeted at a handler, this natural behaviour often meets with a reflex slap on the muzzle. The foal doesn't understand this reaction. It feels unfairly treated and doesn't know how to interpret this training technique. The harmonious, trusting, steady cooperation between horse and handler is upset. The best thing to do is to keep pushing the foal's head aside without comment. If you repeat this neutral, non-aggressive action a few times, the foal will understand that it should not nip or bite its handler. If a foal rears, keep it at a distance using your body language. And never turn your back on it! (see p. 93)

Worming and vaccinations

Foals are wormed for the first time at the age of four to six weeks, and then once every two months, using a suitable wormer. For foals and young horses, we always prefer to use a flavoured paste wormer. Even a shy foal nibbling the tube out of curiosity will eat the paste happily if you can inject it into its mouth quickly enough. Some wormers are not suitable for young foals. Read the instructions thoroughly and, if in doubt, ask your vet. The active ingredient in the product should be different every time. Many species of worm have already developed resistance. To avoid encouraging wormer resistance, give the full dose of active ingredient. Ask your vet or an experienced breeder to estimate your foal's weight.

TIP

Before administering the wormer, feel with your finger to see if there are any bits of feed still in the foal's mouth. Any leftover feed will be spat out together with the wormer paste.

Vaccination against joint ill immediately after birth is no longer practised nowadays. Vaccines for tetanus and influenza/herpes should not be given until the foal is at least six months old. The different active ingredients and vaccinations should be given at separate times. The first basic immunisations are separated by an interval of four to six weeks; after that, a booster is due every six months. These vaccinations are required for horses attending public events. However, they remain controversial.

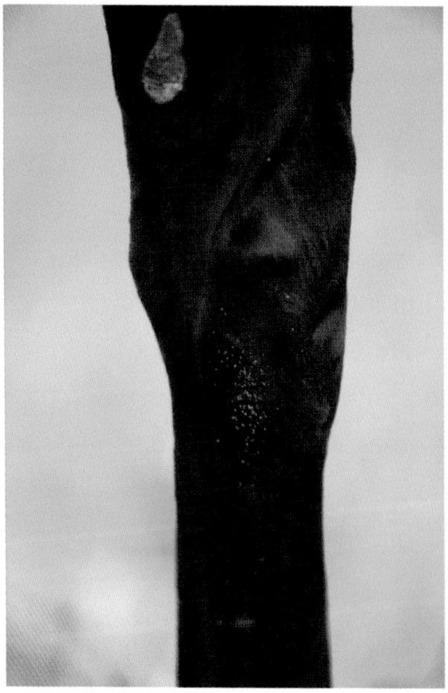

a) Warble flies lay their eggs on the horse's coat. The horse ingests the eggs by nibbling itself or through mutual grooming. b) Daily checks are crucial to make sure that wounds are treated immediately.

The foals are visited daily, including when they are on pasture. Watch carefully to check that they are drinking, eating, happy and active. If a foal seems dull or is not eating, or if the mare's udder is over-full and dripping milk, you have a sick foal on your hands. Call the vet immediately and take the foal's temperature. Diarrhoea in young foals is normal and is associated with the marked changes going on in their intestinal flora. It often coincides with the mare's first heat six to ten days after foaling, and disappears within a few days. In other circumstances, you should inform your vet.

Keep an especially close watch on your foal if it has diarrhoea.

Herbal remedies

If your foal suffers an injury, fall, bump or accident, the authors recommend the following herbal remedy but veterinary advice should always be sought for more serious concerns. Three drops of Arnica C 200 can be given every fifteen minutes initially, until you see an improvement. Slowly reduce the frequency from three times a day to once a day until your foal has completely recovered. Pain in sensitive areas such as the mouth, ears and poll area, the back and the genital region might be treated with Hypericum C 30 (same dosage as above) and your vet can advise on conventional treatments for diarrhoea, which have scientifically proven benefits.

In a case of foal diarrhoea, Podophyllum C 6 can provide relief. Give three drops, three to five times daily, until you see an improvement.

In an emergency

Our herbal remedy tips are not a replacement for veterinary diagnosis and treatment. In cases of deep wounds, diarrhoea or high fever, always call your vet. Foals can deteriorate quickly!

Weaning

Some horse breeders, and pony breeders in particular, do not wean their foals until the worst of the winter is over. Some mares have already anticipated their owner's intentions by then, and have stopped allowing their foals to nurse. Other dams need to be fed generous amounts of supplements to prevent them from losing condition. Especially for these mares, it is kinder

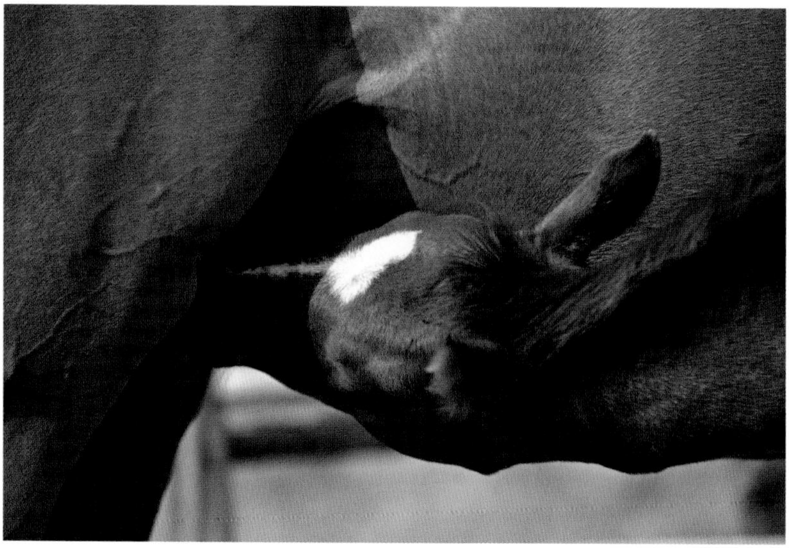

The mother's udder provides the young horse with food and comfort. Both mare and foal cope best with gentle, gradual weaning.

to foal only every other year. Wild horses are similar. On average, their foals nurse for around ten months, sometimes much longer, especially if the mare is not in foal again. Early-born foals are sometimes weaned by their mothers during the grazing season. Weaning by the mare is kindest, not only for the mother's and foal's state of mind, but also for the mare's udder. The mare's milk supply increases constantly over the first three months of the foal's life, and then gradually tails off.

Three-month-old Rhapsody is still too small to be separated from her mother.

Weaning a nursing foal at only three or four months of age is regrettable. Such early weaning is not horse-friendly. The older the foal is when it is separated from its mother, the sooner it will be able to cope physically with the stress associated with weaning. The gentle approach to weaning described below is always preferable to an abrupt separation.

Gentle weaning

If pastures are close to the stable, mares and foals cannot be separated far enough during weaning and will hear each other calling. There is a strong risk that mares or foals will try to reach each other through the fence, illustrating the way that horses feel about weaning imposed on them. So it is important to tackle weaning step-by-step and with a lot of thought.

For gentle weaning, we bring our mares and foals inside overnight starting from around late October, when the nights become colder and the pastures barer. During the day, they are all turned out again. This starts to get them used to living inside. The foals are already big enough to eat forage and concentrates with their mothers. After a few nights, the initial restlessness caused by stabling has subsided. Each foal is given its own extra feeder in the stall. Mares are usually very greedy and will alternate between the two feeders. By watching them, foals learn to eat from the new feeders too. As soon as the foals are eating from the new feeders, the mares are tied up when concentrates are put out. This allows the foals to have their feeders to

Stalls and fences should be sturdy enough to contain a herd of excitable weanlings! But the initial excitement over their new situation will subside within a week.

themselves. Then they are given their future ration of crushed barley, minerals, calcium and foal starter feed, with the amount gradually being increased over three days or so. Once they are finishing the whole ration quickly, it is time to start weaning.

During this time, the mares are fed a low-concentrate, low-energy diet. If you still have oldish hay in good condition, you can feed that as well. The milk supply should start to tail off, but no-one should go hungry!

When the mare herd is brought in for the night, the foals are separated from their mothers. They are put together in a separate stall with a ready supply of forage. Some foals will try to break out, so stalls must be of very solid construction. Some foals eat forage straight away, while others are unsure. Mothers and foals call to each other.

Mares that are already in foal again calm down much more quickly than 'empty' mares. The ones that are having to give up their first foal are especially nervy. The first night is the worst. In the morning the families are reunited, and the foals start to suckle immediately. The mares' udders are full, and the foals need milk, comfort and nearness.

For three to four days, mother and foal are turned out together during the day and separated at night. This gets them used to being apart. Finally, the mares are turned out by themselves one morning on to a heavily grazed pasture with a limited food supply. Keeping them constantly on the move helps to prevent mastitis. During this time, the foals are kept indoors.

In the afternoon, once the mares are back inside, the foals are turned out into a securely fenced paddock. Some foals will try to break through a fence to get to their mothers, so all fences should be as secure as possible and foals should be kept under watch. During the weaning phase it is important for the unsettled foals to have plenty of contact with their owner, as a familiar attachment figure.

After seven to ten days, the situation will have calmed down to the extent that mares and foals are able to eat and sleep properly again, and can be turned out as normal on to paddocks or pasture.

Herbal remedies

To reduce the mare's milk supply, natural healing practitioners recommend giving a teaspoon of dried sage, or a tepid brew of the leaves, in the feed twice a day. To regulate the milk supply, Phytolacca C 30 once a day for three days may be given, but always consult your vet.

If pastures are further away

If the pastures are not parcelled up, you can separate off a small area. For example, you could use a lattice fence that is low enough to allow only the foals to slip underneath. Behind this fence, the foals are fed and encouraged to eat independently. A drawback of this method is that some mares are smart enough to creep under the fence to reach the feed. This may be funny, but it is not practical.

A panel pen composed of mobile fencing elements beside the pasture is a much more practical option. In this pen, the foals are patiently encouraged over several days or weeks to learn the business of eating independently. If the foals are able to eat their future ration without stress, there are various alternatives to the weaning method described above.

To limit the risk of injury, the most important precondition is that mares and foals should not be able to see or hear each other.

Many breeders handle this by taking the mothers out of the herd for an hour or two at first, then overnight, and finally altogether. The foals then

Homeopathy tip

To combat separation anxiety in mares and foals, give each animal five drops of Ignatia C 30 once a day for three days.

An atmosphere of alarm prevails in the weanling group at first, triggering over-reaction among the foals.

take their cues from the remaining herd. The risk of foals trying to break through the fence to find their mothers is low because there is no calling between mare and foal. The success of this weaning method does depend to some extent on the temperament of the foal. But in any case, it is a high-risk method because it is hard to monitor events closely if the pastures are not close to where the keeper or owner is.

However, anything is better than separating mares and foals abruptly.

Moving weanlings to another yard

If you are not planning to rear your little weanling yourself, be on the look-out for a good yard that handles youngstock. Go and visit several yards and keep your eyes open and your wits about you as you take a tour. How are the horses being kept? How large, sound, well-maintained and horse-friendly are the fences, paddocks, stalls, pastures and training areas? How friendly and professional are the people handling the animals? How are the animals

The dynamics of a herd of excited animals are clearly in evidence here. The foals are all bunched together, taking their cues from the older horses in the background.

being fed? What is your impression of their hooves? And above all: would you like to be a horse here?

A move will always entail additional stress for your little weanling. However, you can take steps to minimise that stress. Most importantly, the foal should have been weaned several weeks before the move. This means that he will be more relaxed and able to eat independently on arriving at his new yard. Ideally, he will have had at least his basic tetanus immunisation. He should of course be healthy and free of infectious diseases on the day of the move.

Regular, trusting contact with people is especially important during weaning.

A good youngstock yard makes sure that all foals being entrusted into its care arrive on the same day. They are so busy getting to know each other and their new environment that the pain of separation will fade slightly.

Never add your foal to an existing group of young animals. This can lead to bullying and a high risk of injury.

FOAL VACCINATIONS

Vaccinations should not be given before the foal is six months old. At this age foals are still protected by the antibodies obtained from their mothers' milk, which negate the effect of the vaccine. The basic immunisation is given twice, with an interval of four to six weeks. This means that the foal is at least seven and a half months old before it has completed its basic vaccinations. Correct vaccination at the right time is important. Otherwise, you run the risk of the horse falling seriously ill later on, despite vaccination, because the basic immunisation was not given properly.

The weaned foal – the first winter

Management and feeding in the first winter

AGE
8 to 12 months

The first winter brings drastic changes in the life of the foal, which until now has been closely protected by its mother. This life stage is also important in terms of management and feeding.

The foals are living in their familiar surroundings and still need plenty of exercise outdoors in suitable groups on dry, firm ground enclosed by sturdy fences. The companionship of other horses is important to provide the social contact they need in order to grow up properly. At this age, colts and fillies can still be kept together. An ideal stabling set-up is stalls with direct access to an outdoor paddock and daily exercise on winter pasture.

The foal is still growing quickly in its second six months, so it needs high levels of protein, lysine, minerals and vitamins. Its daily feed rations are divided between at least three meals. Watch carefully to see how the animals interact. They should not be tied up during feeding but animals with feed envy may need to be fed separately.

SUGGESTED RATION FOR WEANLINGS

7–12 months old, medium-sized breeds, in kg/day, weight approx. 265 BM according to Meyer/Coenen (2002). BM (body mass) is the horse's weight in kilograms.

Hay, first cut	2.5 kg
Carrots	2 kg
Oat grains, crushed	2 kg
Foal supplement	1.5 kg
Forage straw	in clean bedding

The foal's skeleton and bone structure

The skeletal system is a framework of around 300 bones which shape and support the horse's body. The animal's mobility and movements depend on the interplay between muscles, tendons, ligaments and joints.

In the embryo phase, the skeleton is initially laid down as cartilage which then gradually turns into bone ('ossifies'). Initial ossification of the skeleton in the horse is already complete by the time the embryo is seven to eight

Fresh air and plenty of exercise are healthy!

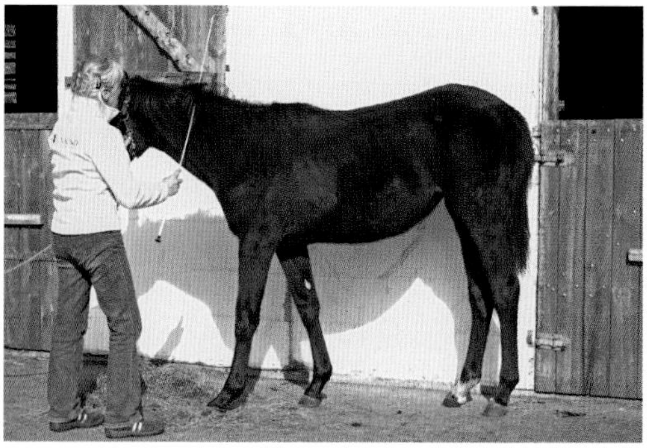

Foals tend to grow unevenly – Universe is too high at the croup.

centimetres long. When a foal is born, its skeletal development is similar to that of a child at puberty.

Longitudinal bone growth occurs at the ends of the long bones ('epiphyses') and in the short bones. During the animal's growth phase, this process ensures that the bones continue to lengthen until the epiphysis closes and the final parts of the cartilage ossify. This means that horses stop growing in height at the age of two and a half to three years. But a horse's skeleton is not fully mature until the bones have stopped growing in thickness. This happens at around the age of five.

Bone tissue consists of organic and inorganic substances. A perfectly developed system of cells and layers called lamellae lends the bone strength and elasticity.

Resilient bone and cartilage can develop only if the young animals have enough exercise. This is backed by scientific evidence, so we strongly recommend that growing animals are allowed as much exercise as possible right from the start. (See Dissertation, Lena Voswinkel, 2009, in Sources.)

WEIGHT DEVELOPMENT	
Birth weight of horse foals	approx. 50–75 kg
Monthly weight gain in the first year	approx. 25 kg
Weight of yearling	approx. 250–300 kg
In the second year	approx. 150 kg weight gain
In the third year	approx. 80–100 kg weight gain

EPIPHYSES

Physiological age of epiphyseal closure (Voswinkel, 2009, modified according to Huskamp et al., 1996)

Epiphyseal plates		*Epiphyseal closure (age in months)*
Ossa digitorum (manus/pedis) (toe bones)		6–9
Ossa metacarpalia III (cannon bones)	distal	9–12
Radius (forearm bone)	proximal	12–18
	distal	24–30
Humerus (arm bone)	proximal	36
	distal	12–18
Tibia (leg bone)	proximal	24–30
	distal	24–30
Tuber olecrani (elbow joint)		24–30
Tuber calcanei (heel joint, part of hock)		24–30

Proximal = growing upwards • **Distal** = growing downwards

The bones' growth regions, called 'epiphyses', begin at the ends of the long bones.

According to T. S. Stashak in *Adam's Lahmheit bei Pferden*, in some cases epiphyseal closure is not complete until the horse is four years old.

However, the horse's skeleton is not fully mature until the second phase of bone development (bone thickness) is complete, at around the age of five.

A sound, trusting relationship between horse and handler.

New calming strategies replace nursing

The newly weaned foal is unsettled and insecure. Measurable stress parameters, such as high cortisol levels, show that the key thing it needs to do to regain its self-confidence is to calm down.

All mammals, including humans, have either inborn or acquired ways to calm themselves down after stress. The foal's main forms of stress relief were nursing and physical contact with its mother. New stress regulation mechanisms, such as regular feeding routines, exercise, contact with other friendly foals within the new group and trusting contact with people, go some way

Stable friendships with peers are vital for all horses, but especially for young ones.

towards replacing previous calming strategies. Nuzzling their friends' bellies, which many foals are seen to do as a replacement activity for nursing, also falls into this category. Even adult horses, especially geldings, will nuzzle into the belly of a four-legged friend – after a painful encounter with an electric fence, for example. Crib biting, another self-calming strategy, can also have its roots in the weaning period and is especially likely to develop after early or sudden weaning.

High stress levels

Throughout its first period of physical and psychological change, the foal's stress levels are high. Depending on how much has changed for the foal and whether it has also had to cope with a change of stable on top of separation from its mother, this period of stress can last for two to six weeks or more. During this time, the young animal is especially prone to infection, injury and colic. We therefore recommend that you avoid additional stress factors such as vaccinations or repeated changes of stable, and especially castration. And avoid starting any new exercises with the foal. Immediately after

Spend time with your weaned foal, watch what it does and enjoy having contact.

weaning is a time of great restlessness in the foal herd. So be especially on your guard when moving among the weanlings.

Spend time with your foal; it is enough to just be there. Your presence establishes a bond between you and your foal. Horses are there for each other too – when dozing, feeding, walking, running or helping with coat care. Notice what makes your foal different from the others, who his friends are, what typifies this special animal.

Only once some time has passed and the foal is more relaxed can you start to step up its training and work on building its relationship with people.

For young mammals, contact with their mother is the first, best and most familiar option they have for calming themselves. If that contact is lost, their stress levels soar. The foal may be unable to deal with excessive stress, even appearing unnaturally calm in some cases. In the wild, this state of 'freezing' or dissociation occurs at times of peak threat when fight or flight is impossible. If this state is not resolved, it can result in psychological trauma. The least stressful situation for a weaned foal is staying in familiar surroundings with other foals that it knows.

Recognising stress

Typical stress responses are 'flight', 'fight', freezing, helplessness and playing up.

In their moving book about free-living mustangs, Hubert and Klein described a two-week-old foal that was separated from its mother for only a few hours but was left physically weakened for several weeks by the stress of separation: 'It was as if all his energy had drained away'.

The foal in its new group

In their new group, the foals need to establish a new ranking order. This process is most peaceful if the new group leader is clearly dominant. The group will be more unsettled if the top-ranking foal has to defend its position because one or more equally self-confident foals want to contest the role. Age often wins the day. Luckily, these types of hierarchy battles are usually confined to feeding times. Groups with an even number of animals are best because they encourage the forming of friendships.

It is sometimes difficult to work with an individual foal within a group. For example, imagine you and an assistant have a foal on a head-collar and lead rope. You are about to work on its feet but immediately find yourselves surrounded by inquisitive foals. A foal that is not at the top of the ranking order cannot stand quietly if it feels hassled by others. Its urge is to give way to higher-ranking foals.

The herd dynamic in new groups is very unsettled at first. Plenty of space and exercise help to resolve skirmishes safely.

This is why we always bring foals into the stable aisle one by one. They know this place, and have been here often with their mothers. Now they have to get to know it again in their mothers' absence (see exercise on p. 79).

An ever-changing selection of safe toys satisfies the young horse's instinct to play.

Foals differ enormously in behaviour depending on their characters. Some foals are completely relaxed and will allow themselves to be haltered, taken out of the group and led away as if they had done it all a hundred times before. But take care: among fifty laid-back foals like these, there is bound to be one that is not as laid-back as it appears. In reality, it has internally switched off to stress and will put up with anything. So keep a close watch on your foals! And, above all, be patient.

Being left alone and 'clinging'

For young foals, clinging to their mothers is age-appropriate behaviour. In the wild, mares often isolate themselves from the herd in order to imprint their newborn foals without being disturbed. Following its mother ensures the foal's survival (see p. 27).

Because they can't follow their mothers after weaning, many newly-weaned foals stick closely to friends of their own age.

If other horses boost the weaned foal's confidence, we include them in our work. Weeks later, we make sure that the foal still has at least visual contact with its friends. In all of our subsequent exercises, we start out either in the large free stall, the paddock or the winter pasture in among the foals, or we use the stable aisle. Sometimes we work with two or three foals in the small paddock beside the stable or winter pasture. This allows them to maintain visual contact with the rest of the herd.

To minimise stress levels, we avoid separating individual animals from the others completely. This would make them too anxious and an anxious animal won't learn anything. When the foals have matured into two- or three-year-olds, there is still enough time to get them used to being alone in the indoor school or out on a walk. Mares tend to mature later in this respect, often sticking very closely to their friends. Even as two-year-olds, stallions and geldings will sometimes stay calm and relaxed when being led into the school by themselves. This is in the nature of horses. Mares are never found on their own in the wild, unlike stallions, although the latter tend to stay with the group until the age of two.

None of our horses are clingy, they have all learned to work on their own – but none of them were made to learn this exercise as a foal if they weren't ready for it.

Our method of working in small groups is age-appropriate and horse-friendly. It promotes the young horse's trust in you as an attachment figure. It also trains your antennae to pick up the mood of the herd, and your ability to be one hundred per cent present at all times when working with your young horse. This is a skill that marks out a true horse lover just as much as treating the horse fairly.

And we are not just concerned with horse-friendly rearing methods, but also with our own safety when leading or riding a horse.

Following or 'clinging' is part of a foal's natural behaviour.

DO YOU UNDERSTAND YOUR FOAL?

When we suggest spending time with your foal, we are not saying that you have to stick to it like glue. Take your camera along, look up from time to time, and watch your foal or take some photos. Familiarise yourself with his facial expressions, his role within the herd, the way he moves, his habits and his patterns of behaviour. By doing this, you will avoid the later risk of mistaking anxiety for stubbornness or aggression and treating your horse unfairly. You will recognise tight nostrils and lips, narrowed eyes and ears pointing slightly out to the sides as a 'stress face' and be able to respond with appropriate patience.

The 'grooming face'

The 'stress face'

The 'playful face'

The 'inquisitive face'

Reaching an understanding with your weanling

Once your weanling has 'processed' the separation from his mother and is feeling more relaxed in his new herd, you can start to engage with him more actively. But before trying to teach him anything you need to consider the basic rules of communication.

Talking to horses

As far as words are concerned, it is a good idea to decide as early as possible which commands you plan to use. It doesn't matter whether you say 'stand', 'halt', 'stop' or 'whoa' when you want your horse to stop. The main thing is

Words are not necessary for a meaningful dialogue between owner and horse. Marietta and Liberty understand each other through body language alone.

Universe is still eyeing the cones nervously but Ute's posture is active and focused on moving forward. So Universe allows herself to be led.

to be consistent, because chopping and changing between different commands for the same thing causes confusion.

It can be helpful to explain a situation to a horse in the same focused and detailed way that you would use with a person. As you do so, keep a clear idea in your own mind of what you want from your horse. This will help to get the right message across. Use words as consciously and deliberately as possible, either singly or in short phrases.

Body language

The most important aspect of communicating with a horse is our own clear, conscious body language. You can encourage shy foals to relax by engaging in a leisure activity or repairing a fence in their vicinity. They will gradually get bolder and come closer of their own accord in order to sniff you. Adopting a slightly hunched posture with rounded shoulders, and lowering your gaze, can also help to relax them. But if you straighten up suddenly the foal will interpret this as an attack and take flight. Practise both postures in

Holding a feed dish low down encourages Princess's relaxed posture. Chewing the oats also has a calming effect.

the mirror and observe their effect. Your posture as you walk can also have quite different effects.

Keep a general lookout but focus your attention on your foal and its surroundings. Use gestures deliberately and sparingly. Move in a relaxed way. All of this mimics the behaviour of a good lead mare.

Praise and rewards

Praise is an important signal for our horses. Get into the habit of praising your horse whenever things are going well. Give praise by tickling, stroking or feeding, or with a soft or enthusiastic 'good boy/girl' or 'well done'. Young horses need to learn our language, our aids, our wishes, before anything else. By reinforcing what they have learned, we make them more trusting and more secure in themselves. Praise every response, even a hesitant one, and praise immediately. On the one hand, this teaches you to watch carefully. On the other, it motivates shy horses to work with you happily.

The end of the work, or at least a short break, also comes under the category of 'praise' (see also p. 101).

Punishment

TIP

There are some days when you shouldn't attempt to do anything with your foal. On days like these, just stay nearby and watch it.

But what to do if your young horse's behaviour falls far short of praiseworthy? What if it rears, kicks out at you or bites?

A horse is a large animal. Fully grown, it weighs around 500 kilos and can easily injure a human. This may be why we tend to respond in less than clever ways when a horse feels threatened.

In most cases, the behaviour we want to punish is an activated flight instinct. Harsh treatment reinforces rather than reduces your horse's desire to flee. We advocate the path of trust. This creates a sound, reliable basis

Like her mother, grandmother and all of her siblings, Universe responds to stress by vigorously pawing the ground. She is told 'no' but is also stroked with the whip to reduce the tension.

for your later work together. The time that you invest early in a horse's life in getting it used to procedures such as giving a hoof creates trust, and also consolidates and strengthens your relationship. Horses that have learned to fear people are always tense on the inside, which makes them unreliable.

On the other hand, there are some situations that can be clarified by a short, sharp 'no' or by standing up straight and making yourself look bigger. If you punish your horse with a loud 'no' or 'shhh', do so immediately after its unwanted behaviour. Similarly, a horse punishes another herd member at the precise moment when it misbehaves. Here too, it is worthwhile observing what mares do. Mares are quick to act if a foal oversteps the mark, and they don't bear grudges.

And any time you do have a mishap, always ask yourself how the animal's unwanted behaviour could be avoided the next time, and whether your punishment was justified. Often, the fault lies with the handler.

WHAT CAN GO WRONG?

- Did you practise for too long?
- Perhaps you didn't prepare the young animal properly for a new situation?
- Was it disturbed by a change in its daily routine?
- Were there any disruptive factors (general hustle and bustle, a new stablemate, bad weather, etc.)?
- Were your signals clear enough?
- Did you have a bad day yourself?

Exercises with the yearling

Once you have established a trusting relationship with your foal, you can try some exercises. But always allow yourself plenty of time and be patient. The main aim is to get your weanling used to certain things and situations. Not to get everything perfect.

Leading

For leading, you need a properly fitting adjustable halter or head-collar, and a lead rope. It is nicest for the foal if you use an ordinary rope. We lead very young animals by passing a long rope through the head-collar and holding on to both ends. We also recommend this on unfenced terrain. In an emergency, you can let go and the animal won't tread on the rope. Another useful option later on is a rope with quick-release panic clips. Only with two-year-olds that are already used to being led do we work by the book, i.e. with a rope tight around the nose. And we do this only when working on fenced ground. It is simply too risky otherwise: a young horse might pull free and sustain serious injury. If a horse did pull free and somehow managed to tread on the lead rope at full gallop there could be serious consequences, including dislocation of the neck.

For sensitive young animals, knotted head-collars are just as unpleasant to wear as head-collars with a double nose band. They press on highly

Foals learn best in a group.

Note the relaxed postures of handler and horse.

sensitive nerve points on the face and often sit too low over the nostrils, restricting the nasal passages and obstructing the animal's breathing.

A thick rope without clips can be used around the chest in addition to the lead rope if your young pupil has a tendency to take off.

When leading, we also work with a stiff dressage whip, which we use as an extension of our arm. The whip allows us to stroke overexcited young horses, which calms them down. It is also useful for keeping some distance between you if the foal tries to nibble your arm.

Sturdy footwear and thick gloves are recommended when leading.

An environment free of disruptions is important for any new exercise.

And when working with young animals we always do so in the company of other horses (see p. 66).

Posture during leading

When leading your foal, make sure to stay in front of its shoulder. If you fall too far back, it feels as if you are pushing. It is also hard to control your foal from behind. Young animals tend to walk on the forehand, often causing them to rush. Bends, turns, obstacles and occasional halts can help to curb the pace slightly. Feeding slows them down too. In general, however, you should initially adapt your own pace to that of your foal, and move as smoothly as possible. You are not aiming to get your horse moving in slow motion. Instead, you want to get it as balanced as possible from the ground up. Later on, during breaking in, there will come a stage where the young horse is inclined to bolt forward, putting too much weight on the forehand. But good-quality ground work before breaking in can make this stage much shorter!

When you start working with your foal, he is both physically awkward and psychologically out of balance. His flight instinct is activated. Body work, feed, calming words and a relaxed posture help him to remember that you are someone he trusts.

Gradually, your foal will recover his physical and psychological balance. He will become more self-assured, more aware of his body and more trusting. Age-appropriate ground work using obstacles will teach him to shift his weight more to his hind end. This will also reduce his tendency to rush.

EXERCISE 1: TOUCHING

Aim of the exercise

To build trust, offer comfort and reduce stress.

What you need to know

Give yourself and your weaned foal the pleasure of regular contact through touch. Enjoyable touch is especially important at this time, as it promotes the production of oxytocin, the anti-stress hormone. Sympathetic owners often have high stress levels themselves after weaning, so by stroking, tickling or massaging your foal you are also doing yourself some good. When you stroke the foal's body with your hand or the whip, make sure it enjoys your touch. Find out which touches it likes. It should not simply tolerate things; it should enjoy being with you.

If working in the stable, make sure there is no risk of being barged by other weanlings. The process of establishing a ranking order among the foals is already in full swing.

Deliberate, gentle touches are calming and build trust.

How to approach the exercise

In the first few weeks after weaning, the main purpose of touching the foal is to offer comfort, to strengthen your bond and to build trust. Scratch the foal in its favourite spots, building in some TTouches as described earlier.

After a few weeks, once the foal has come to terms with weaning, include other areas of the body and stroking with the whip as described below.

EXERCISE 2: STROKING WITH THE WHIP

Aim of the exercise

Stroking with the whip is an excellent way to get young and old horses alike used to being touched in sensitive areas, for example when you need to treat a wound. It also calms and relaxes the horse in unfamiliar or stressful situations such as vaccinations, the taking of swabs or a first attempt at saddling.

Marietta combines stroking with tickling.

How to approach the exercise

Stroke down the neck, over the chest and forelegs, to the hooves. The back and hind legs can also be stroked with the whip. If the foal stands nicely and stays relaxed, you can include his belly in your stroking programme. Ideally, your weanling is already familiar with stroking from his time as a nursing foal. But if you need to get him used to it first, turn the whip around and hold it with the thick handle projecting a little from your hand as you use it to stroke the foal's neck.

Gradually let the whip slide out more from your hand, and turn it round the other way. Stroking with the whip is a fairly casual touch, like the mother's rhythmic licking or the drifting of her tail across the foal's coat. Breathe in and out slowly and evenly, and make sure you are standing comfortably.

Stroking with the whip makes Universe aware of her hindquarters. She follows the touch attentively.

EXERCISE 3: PREPARING FOR TYING UP LATER ON

Aim of the exercise

Allowing itself to be tied up is a basic exercise, but doesn't really need to be mastered until the horse is three years old. In preparation for this, your foal can learn to control his impulses briefly.

What you need to know

We never tie our young weanlings. The risk of injury is huge if a panicked foal gets tangled up in the rope and starts to struggle. The foal's immature bones and ligaments are very vulnerable, and not just in the neck area! As a prerequisite, your foal should have done the first part of this exercise (p. 42) several times and should stand relaxed when haltered.

What you need

Foal head-collar, whip and a rope around three metres long.

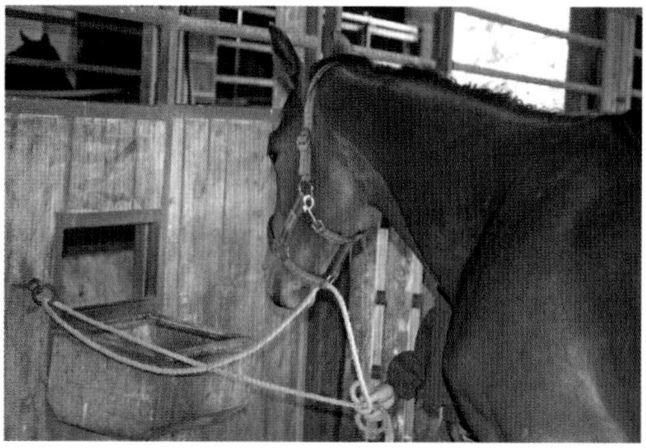 

a) The rope is passed through the head-collar ring and the tie ring; both ends are held in one hand. b) One end of the rope is tied to the head-collar using a quick release knot. If necessary, the foal is given more space immediately.

How to approach the exercise

Pass a long rope through the bottom ring on the head-collar and hold both ends of the rope in one hand. Now hold the foal still for a few moments. It should notice the restraint but not be afraid. If the foal is unsettled, allow him some more rope. If he panics, release him immediately. Usually, you only need to 'give' slightly to calm the panic that some foals feel on being restrained. With time, foals learn that there is nothing to fear in being restrained. If the foal is relaxed, hold one end of the rope in your hand and tie the other end to the head-collar using a quick release knot.

EXERCISE 4: FETCHING INTO THE STABLE AISLE

Aim of the exercise

Being separated from the group is another important step in a weanling's training once it is used to its new situation without its mother. The stable aisle is a good place for hoof work and your first leading exercises.

What you need to know

As a rule, foals will happily approach and make contact with unfamiliar horses in the stable aisle. This is a real mini adventure for the young ones, and provides a good opportunity to study stress faces and foal chewing! The situation is also exciting for the foals left behind. The first excursion lasts for around ten minutes. Nervous foals that are not keen on being led or haltered are happy to return to their stall afterwards.

More laid-back foals really enjoy their first excursion into the stable aisle and would happily spend the whole day there. If necessary, your assistant can help to push the foal back into its stall.

What you need

A properly fitting head-collar, a lead rope, a loose rope around the foal's neck, a dish of oats and an assistant to keep the whole foal herd from bursting out of the stall all at once.

How to approach the exercise

Put on the head-collar and drape the loose rope around the foal's neck. Then lead it out of the stall on a lead rope. If it doesn't want to leave its herd mates, tempt it with the oats and give it a push from behind if necessary. Some foals rush forwards straight away, while others stop to take in their surroundings. Our job is to calm the foal by stroking it with the whip and offering the dish of oats. We adapt our own pace to that of the foal as we walk along the aisle, halting briefly now and again, stroking all the time, before returning the foal to its stall. After repeating this exercise a few times, we can start to halt in the aisle for longer, in order to stroke the foal with the whip, pick up a hoof, etc. Lower-ranking foals usually stand better in the aisle than in the stall.

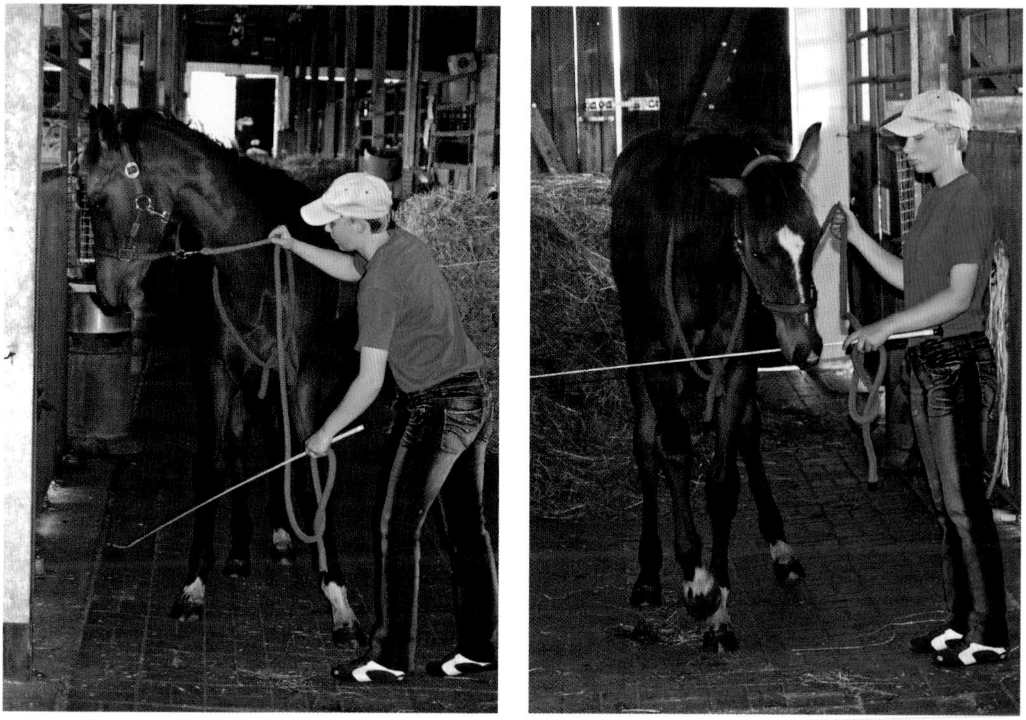

a) Marietta strokes the excitable Starlight with the whip. b) Stroking starts to take effect. Starlight lowers her head and concentrates on Marietta.

EXERCISE 5: GIVING A HOOF

Before training your foal to give a hoof, use calm, even stroking by whip and by hand to get it accustomed to touches on its fore legs, hind legs and hooves. On occasion, you can also touch or massage its legs without lifting a hoof.

Giving a hoof demonstrates enormous trust on the part of a flight animal. Stay patient when you do the exercise and expect the same patience from your farrier. Put off the first farrier's appointment if you sense that your foal might still be too nervous due to weaning, a change of stable or a new group composition, or if you think the farrier might try to control the foal by force. Ask around for recommendations on a horse-friendly farrier!

The front hooves

Before lifting a hoof, take a look at how your foal is standing. Make it a habit to check its stance before you start to change anything.

To practise giving a front hoof, stand at the foal's shoulder facing its tail. Stroke down the shoulder towards the hoof. Note how the foal reacts. Does it stand still? Does it play up? Does it nibble your jacket?

Now grasp the foal's fetlock joint and lift. Use the signal word that you plan to use in the future for hoof-giving. If your foal doesn't understand what you want, scratch upwards along the back of the cannon bone.

Universe carefully tracks what Beate is doing with her legs. Stroking and TTouches on her legs have prepared the foal well for giving a hoof – she is already lifting her leg.

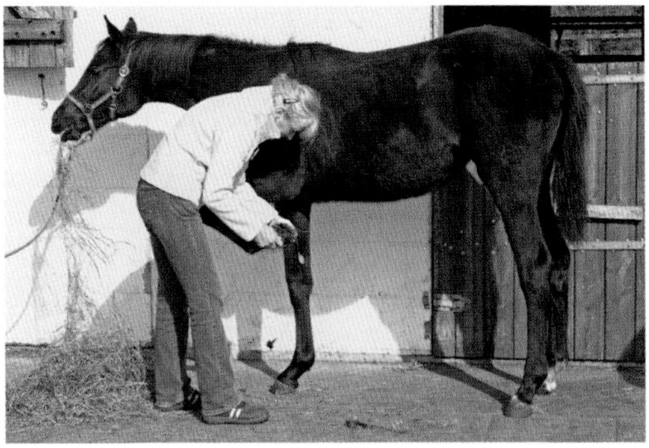

Universe is already giving a hoof for Beate like a pro; it helps that she is being held and has hay to munch on.

Beate picks out Universe's foot, holding the hoof nice and low. This is more comfortable for the young horse.

A slight shift of the foal's weight is enough for a first time. End the exercise with plenty of praise and enthusiasm. This exercise will gradually get the foal used to giving its front hoof briefly.

Many horse people suggest leaning against the horse to make it lift its leg. We do not recommend this. If you lean against a foal to force it to lift a hoof, it will lose its balance, feel insecure and start to fidget.

The rear hooves

Some weanlings are not keen to give a rear hoof. It makes them feel insecure, and they protest. If you have an animal like this, get it used to having its hindquarters touched by stroking with the whip.

Some feed offered by a second person can help to relax the foal. Like us, horses can't chew and hold their breath at the same time! Hold the hoof only briefly at first; ideally, put it down before the foal begins to pull away.

Does your foal give both hooves equally well?

Preparing for the farrier

What you need to know

Malpositions can be corrected, but only up to the age of 18 months. So make sure you introduce your foal to an experienced, patient farrier. Serious deformities like club foot need intensive corrective work right from the start!

Important

Find yourself a patient farrier. The foal's first experiences with the farrier will determine how he behaves with every farrier in future!

Besides leading the foal into the stable aisle, halt it there and get it accustomed to the movements and noises that the farrier makes.

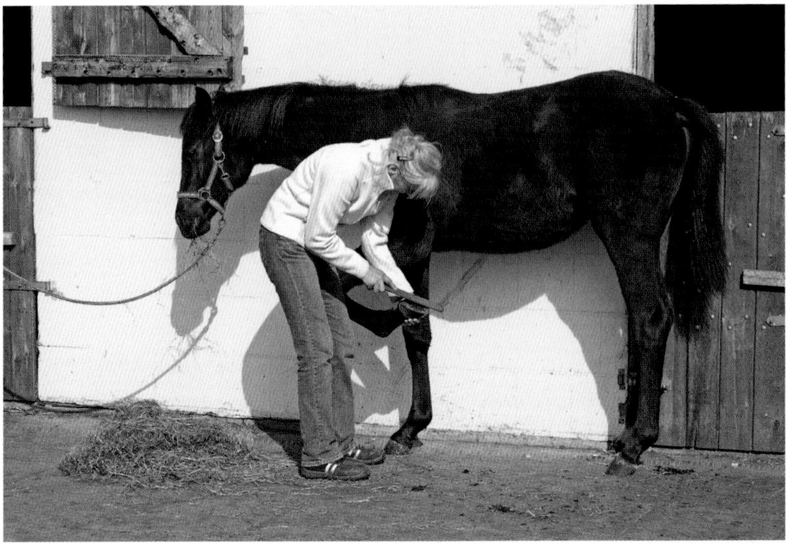

To file Universe's hoof, Beate holds the foot at the same height that the farrier would do. The foal's ears are pricked towards the unusual goings-on.

The yearling on summer pasture – the second summer

Management and feeding of yearlings

AGE

12 to 18 months

From the turn of the year, the weaned foals are described as yearlings and are classed as one-year-olds – even if they were not born until July.

As soon as the weather and grass growth allow, the yearlings are turned out on to generous, securely fenced pastures from mid-April until around mid-November.

If not done already, colts and fillies are separated. Fillies go into a mixed mare herd with animals of all ages, while colts are separated by age into stallion herds. Mixing one-year-old and two-year-old colts, for example, leads to a high risk of injury. Play fights between stallions can be very rough. The older animals are larger and heavier and can easily cause serious spinal injuries in the younger ones (see p. 98).

On the other hand, a calm and friendly older gelding can be a useful member of the group. Many breeders and rearers report positive experiences with older role models. For example, a retired sport horse can be given a responsible role as a member of a young stallion herd.

This is the start of a more relaxed time for the breeder, with the stalls empty and the horses eating outside. Fences and drinkers should be checked every day, and the yearlings still need daily contact with people. Daily health checks and occasional trips out of the pasture while maintaining visual contact with the other horses are important in order to reinforce lessons already learned. Summer may be a quieter time but you still need to give wormers, treat disease or injury and attend to hooves.

29-year-old foundation mare Radschendra is the former matriarch of the mare herd and is still widely respected.

Feeding

In the spring, the horses are given high-fibre, low-protein supplementary feeds such as straw or last year's hay. From late summer onwards, their energy balance is maintained by feeding concentrates.

To prevent health problems, horses of all ages are given limited access to pasture, with the number of hours on grass being gradually increased. While housed in the stable, they are fed concentrates and forage in addition to fresh grass. The rule is as follows: forage in the morning before turnout, concentrates in the evening after bringing in.

FAMILY TIES

With each new year, it is exciting to see how the various family members find each other again in the mare herd. For example, if a breeding mare doesn't have a foal at foot this year, it only takes a few days for last year's foal to find her. The foal may not be nursing any more, but it is back together with its mother. Things are slightly different if the mother has a new foal. In this case, the yearling has to make do with its friends.

One of our brood mares, the youngest daughter of our 29-year-old foundation mare, didn't have a foal for three years. For those three years she looked after her mother, watching over her and sticking by her side when there was unrest in the herd. This year, she has another foal of her own which is now protected by two mares: its mother and its grandmother.

Mother and daughter indulging in a spot of coat care.

Healthy yearlings that are properly developed for their age are allowed to be a bit rough around the edges – ideally, they shouldn't look too 'proper'!

After a week or so of daytime turnout, the yearlings can be left out all day and all night if the weather is mild. In the first months of spring, fresh pasture grass meets all of their energy needs. In dry years, however, the energy and protein supplied by the grass may not be sufficient. If this is the case, supplementary feeding is used.

Mineral feeds should really be given according to the animals' needs, but mineral licks are convenient and have proved their worth. Horses should always have access to a salt lick. If the supply of minerals is inadequate, the body makes up for deficiencies by leaching minerals from the bones.

We therefore advise against intensive rearing and early use. Excessive feeding is very high-risk. The muscles can be over-pumped, and the young animal looks disproportionately 'ready' even though its bones, tendons and ligaments have not kept pace. There is a proven link between excessive feeding and bone chips. Steady growth throughout the rearing stage is more normal for the animal, and healthier. If you want to present your horse successfully for stallion licensing and performance tests, on the other hand, it is essential to adjust your grazing, feeding and training in line with generally accepted practice.

But you can still exercise your horse freely in paddocks and not overdo its feeding and training.

Sexual maturity

Colt foals become **sexually mature** at 12–20 months, and fillies at 12–18 months. However, these are simply averages based on practical experience. Biological sexual maturity, i.e. the age at which a horse is able to reproduce, is different from breeding maturity. This is the age at which a horse is allowed to reproduce. For genuine wild horses, therefore, breeding maturity is meaningless.

In domesticated horses, **breeding maturity** is determined by the breeder. Responsible breeders do not mate their mares before the age of three. After licensing, stallions are generally first used for service in the following spring, when they are three years old.

At our stud in Heidekaten, Germany, we rear colts and fillies together. We aim to allow them to behave like proper foals until they are turned out to pasture, without hormones getting in the way. But if the group includes even one early-maturing stallion, the animals will have to be separated by gender.

One year, we had seven fillies and only one colt foal but he proved to be really early maturing. At the age of 10 months, he started to mount his friends. Some of the fillies in question were already over 12 months old and had visibly come into heat. So we had to separate the colt from the fillies. We bought in two other stallions of the same age, and the young colt was then raised with them.

Young stallions are not necessarily productive when they first become capable of mating. Sometimes they can serve a mare without her becoming pregnant. But even 'unproductive' matings like these entail a lot of stress, restlessness and risk of injury for the herd. The young mares kick out at the stallion, and the horses lose weight because they are not able to eat in peace. In the wild, the alpha stallion would not allow yearlings to mate. And nor should we allow it!

If an actual mating takes place and a young mare becomes pregnant at only one or two years old, she should be monitored as closely as possible

The harmony of a coherent mare herd living in natural conditions also transmits to the people who handle or even just observe the horses.

Young stallions need other stallions to play with. Rearing, biting, racing and barging all feature on their daily timetable . . .

during her pregnancy and looked after extremely carefully. The foaling itself should take place in a veterinary clinic.

The mare's immature pelvic organs and birth canal can be easily damaged during foaling. She may even suffer internal bleeding or bruising and damage to the lumbosacral plexus, making her unable to stand after the birth.

Handling the young stallion

If you own a young stallion, you will inevitably have to deal with the issues described below. Nipping, rearing and play fighting are simply part of a young stallion's daily routine; they are all part of the animal's natural behaviour.

Help, he's barging and nipping

It is entirely normal for young stallions to nip, rear and barge. They practise these typical male activities for several hours every day. Things become tricky when they start to practise on us. If they are doing it because they don't have a male companion of their own age, you need to introduce at least one other stallion. But if they are already growing up with other stallions, they need to learn that people are not suitable playmates for their macho games.

A slap on the muzzle is often used as a training method, but we strongly advise against this. Nipping or barging is your stallion's way of inviting you to play. If you punish him for it, you will lose his trust. To deal with a nippy horse, we push his head aside without comment or with a firm but affectionate 'stop it'. After this has been repeated a few times, all young stallions will get the message.

Rearing or kicking

Rearing is another natural play behaviour for a stallion. All healthy, vigorous young stallions will rear now and then.

. . . and also train strength, agility, coordination and stamina.

Never turn your back on your young stallion, because this can trigger his impulse to rear. When releasing him on pasture, always turn his head towards you so that his hindquarters point away from you before letting him off the rope.

Take a few steps backwards if a young stallion rears in front of you. Growl at him gently, so that he understands you regard his behaviour as inappropriate. You can also show your displeasure through your body language: stand up straight as you scold him!

Young stallions often kick each other, but they will also kick out at people. Here again, we recommend punishing the stallion with words rather than with a slap or a blow. This will help him to regulate his behaviour without losing his trust in you.

Castration

We recommend that young stallions are not castrated until they are at least 18 months old. In smaller herds, however, it is often not possible to keep small groups of stallions. In this case, a good option is to move the colt

INJURIES IN MIXED-AGE STALLION HERDS ON PASTURE

Strength and power imbalances and continual restructuring of the ranking order can lead to violent fights between stallions, resulting in injury. The injury most familiar to the non-professional is ataxia, in this case caused by a blockage of the second and/or fifth cervical vertebra. In ataxia, this blockage is so severe as to crush the spinal cord, causing uncoordinated movements. If a kick from a stronger stallion connects with the shoulder blade or hip bone, the outcome can be a fracture.

Sometimes, in the wake of a tussle, you may see a stallion standing on three legs and dragging a foreleg. This is a foreleg ataxia, caused by a blow to a nerve in the shoulder. Minor kick injuries, and ganglions or chips, are not serious provided they don't affect the course of a tendon or the function of a joint.

All other types of blockage, as described on p. 43 under 'Taking care when leading and tying up', can also happen in the paddock.

Gina Seydel, physiotherapist and equine osteopath (DIPO)

foal to a different youngstock yard where it can grow up with other young stallions. Ideally, visit a number of yards and choose the one that treats its horses best (p. 56).

As a general rule, castration is carried out in either spring or autumn, when the nights are cooler and there are not so many flying insects around. However, conditions should not be really cold, stormy or rainy.

Conditions on pasture are much more sterile than in the stable. And the horse's constant movement on pasture can help wound secretions to drain properly. However, feeding is a problem. The horses will be given a general anaesthetic, so they should really be fasted. If you don't want to fit a muzzle to avoid stressing the animal, you can at least move it to a relatively heavily grazed pasture before carrying out the castration.

For more about stallions, see from p. 90.

Mixed herds including yearlings

You should visit and check your horse every day, in summer as in winter. It is already used to being led into the stable aisle alone, so there is no reason not to work in the small paddock beside its pasture. Here, you can practise touching and easy ground work once a week or so, to consolidate what it has already learned. If you remove it from the group entirely, without visual contact or closeness with the other horses, it will become agitated, making it almost impossible to work with successfully. Leave this stage of its training until the following summer. It introduces unnecessary restlessness into the herd dynamic.

Quite apart from the risk of injury, your horse will come to associate your contact and its training with stress and disturbance. This is not a desirable outcome.

The herd is your yearling's main classroom in the summer months!

Herbal remedies: tips for castration

Natural healing practitioners recommend that the number one wound treatment, Arnica, can be administered in a high potency of C 200 several times a day or every hour for the first day, then three times a day for the next three days, and then once a day until the wound has healed.

For complicated castrations or the removal of an abdominal testis, the same doses of Bellis perennis rather than Arnica is recommended.

For wound healing disorders, Silicea C 30 is recommended, and for abdominal swelling, Apis C 30, both twice a day, until improvement is seen.

Always consult your vet.

In the herd

Bee keepers often refer to a colony of bees as 'the hive', treating them as a unit. A herd of horses can also be regarded as a single organism. Each horse has its own place within the herd, which is held together by a web of relationships between horses. Once a ranking order has been established, the herd functions smoothly. It has its own laws, play times and rest times, and its favourite places to doze in.

Each horse responds individually to changes in the weather, daily and annual rhythms, restlessness, disease, birth or death. But the herd also responds as a whole. People nowadays often lack an instinct for the herd organism. We only see the individual horse: our horse.

It may have behaved politely in our exercises just a few minutes ago, yet here it is galloping around like crazy on fresh pasture with its herd mates.

Developing a feel for the herd mentality can be challenging. But for our horse, the herd is essential.

From yearling to two-year-old – the second winter

Management and feeding

AGE
18 to 24 months

The yearlings are brought back into the stable in autumn. Precisely when depends on the weather conditions. The ideal set-up for all horses, including two-year-olds, is stabling in loose boxes with access to an outdoor paddock and winter pasture to guarantee plenty of fresh air, light, companionship and exercise. But even in other set-ups, two-year-olds, like all horses, need several hours of daily exercise in the group (see p. 61). It is also important to have trusting daily contact with the people caring for them.

The horses' daily routine now includes exercises that are designed to consolidate lessons already learned. These exercises are generally more demanding than those in the first winter. One exercise that can be incorporated easily into their daily routine is leading to and from the paddock. This allows you to practise haltering, attaching a rope to the head-collar and waiting politely until the gate has been opened. Lead from the right and left alternately.

The feed ration consists mainly of good-quality hay, haylage and a concentrate supplement. The young horse has already completed its main growth phase and the amount of foal starter feed can be reduced.

The ration should be divided into at least three portions per day.

TWO SUGGESTED RATIONS

In kg/day for yearlings 18–24 months old, medium-sized breeds, weight approx. 400 BW according to Meyer/Coenen (2002). BW stands for body weight in kilograms.

Hay, first cut	3.5 kg	2 kg
Wilted haylage		5 kg
Wheat straw	1.5 kg	
Fodder beet	4 kg	
Crushed oats	1.2 kg	2 kg
Supplementary feed for foals	1.5 kg	1 kg

The daily rations given in the table should be divided into three portions.

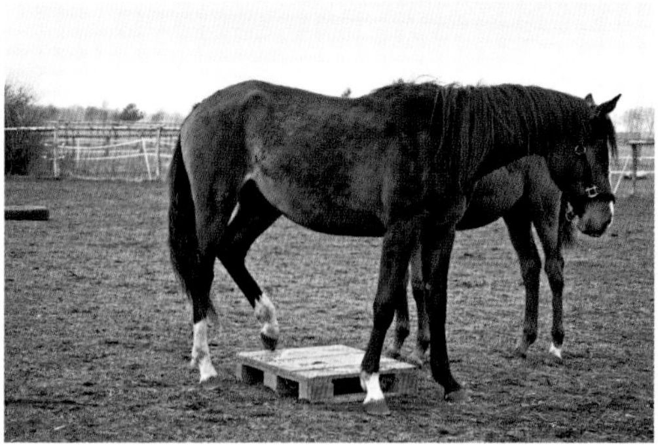

Horses enjoy exploring together. Toys, such as this pallet reinforced with floorboards, should be safe.

How do horses learn?

Horses learn in various ways. They are most receptive to learning and open to new things when in the company of experienced older horses. This makes them feel safe. They can observe free from stress and learn by watching.

PAYING CLOSE ATTENTION

My 10-year-old trotter Manolo and I used to practise bowing in the pasture in among the other horses. One day, I was about to reward Manolo for maintaining his bow for a particularly long time when I saw a second open mouth appear alongside Manolo's head. Our Trakehner Kofi, a two-year-old at the time, had followed each part of the lesson attentively and had copied Manolo's bow right by his side. Now he was justifiably demanding his reward!

But take care: horses don't only watch and learn desirable outcomes. From his friend Kofi, our young Haflinger Shoshoni acquired an unwillingness to ride on a particular stretch of moorland. Kofi had spooked there once when he had seen a giant poplar tree being felled. Although Shoshoni wasn't there at the time, he had picked up the idea that there was something horrible on the moor and it was best avoided.

Learning by imitation

Even very young foals learn by observing other horses' behaviour. This type of learning never ends. You can turn it to your advantage by using experienced horses as training partners alongside 'green' animals during ground work, trailer training or simply going for a walk. Make sure these old hands really are as safe and brave as you would like your own horse to be, because your horse will pick up not only relaxed attitudes but other behaviours as well.

Learning by positive reinforcement

Many dog trainers, and increasingly horse trainers too, reward with treats or deliberate positive reinforcement. Apart from food, other forms of positive reinforcement include breaks, being allowed to run free, a favourite exercise, enjoyable physical contact or the end of the work. Food has always

Rhapsody looks on with interest as Marietta puts a head-collar on her friend. Pandsy stays relaxed, so now Rhapsody too sees a head-collar as unthreatening.

A break, some food, stroking with the whip, a gentle caress under the jaw or a scratch on the mane – you can never give enough praise.

been used as a learning aid in the training of circus horses. Clicker training is an extremely effective method devised by the American, Karen Pryor, who developed it for work with dolphins. Treats have a highly motivating effect on horses and also train our own eye, both to break tasks down into chunks and to identify small milestones that need to be reinforced. We use food to calm young horses, and to speed up the learning process in older animals. If you want to try clicker training, a method we enjoy and recommend, we suggest you seek instruction from a specialist with relevant experience.

More ambitious ground work

In the following exercises which you can do with your young horse, previous lessons are consolidated and the degree of difficulty is stepped up slightly. As before, these exercises train communication between you and your horse. They are also a playful way to train body image.

EXERCISE 1: HALTING

Aim of the exercise

This is a basic exercise. It refines communication between handler and horse; the horse learns to focus on the handler and to control its impulses.

How to approach the exercise

To halt your young horse, start by correcting your own body position if necessary. You should be ahead of a line level with the horse's eye. Tug on the rope, direct the whip in front of the horse's chest or nose and give a verbal signal. Keep walking until your horse halts. If you stop immediately after giving your 'whoa' or 'halt' signal, you will end up behind the horse's eye. In most cases, the horse will then make a small left-hand turn around you before coming to a stop. It is better to say 'whoa' or 'halt' and then count mentally to around 20 before calmly halting. With young horses, you may even have to count a bit higher!

Don't forget to praise your young horse immediately as soon as it responds. Even a hesitant slowing down, for example, should prompt enthusiastic praise.

Remember to practise leading from both sides.

A short joint training session.

Ground work can now become slightly more ambitious. This exercise involves halting in a triangle of poles.

Beate gradually leads Princess out of the triangle, halting her repeatedly.

EXERCISE 2: STANDING STILL

Aim of the exercise

This is a basic exercise in a horse's initial training. The horse learns to control itself. Its internal and external sense of balance, and its patience, are improved. The handler learns to sense when the horse is about to move and pre-empts the animal with a command to walk on.

What you need

Properly fitting head-collar, rope with quick-release panic clips, whip, possibly some jump stands, poles and foam tubes.

What you need to know

This exercise can tell you a lot about your yearling's character and temperament. Some horses find it easy to stand still, while others last for only a few seconds before they have an urge to move. Make allowances for your horse's temperament while bearing the aim of the exercise in mind.

How to approach the exercise

Practise in a familiar environment such as the stable aisle. Give the signal to halt, and the horse should halt. Now give the command to stand still, for example 'stand'. Stroke the horse with the whip. While the horse is standing still, you can touch it all over, pick up its hooves or groom it. At the end of the exercise, praise your horse and ask it to walk on before it moves off of its own accord.

Follow-up exercise

If the exercise goes well in the stable aisle, you can also try it within the confines of the indoor school or an outdoor arena, but always in the company of other horses. Using poles, jump stands or foam tubes to limit the space will make it easier for the yearling to stand still. Standing as described on p. 130 is a more advanced exercise.

Remember to practise leading from both sides.

EXERCISE 3: WALKING ON

Aim of the exercise

This is a basic exercise. The horse learns to concentrate and to respond to the tiny signs given by the handler. This exercise promotes its sense of balance.

What you need to know

Ask your horse to walk on before it decides to set off for itself. Unbalanced young horses with a lot of weight on the forehand tend to set off at a rush.

What you need

Properly fitting head-collar, rope with quick-release panic clips, whip.

A labyrinth of two intersecting Us or Fs composed of six to eight poles. The horse is halted after each turn inside the labyrinth.

How to approach the exercise

Take care to act, rather than react. For example, if you feel that your yearling is no longer standing still but is impatient to get going, command it to walk on, signal 'forwards' with the whip and set off determinedly. But if your yearling is more reticent, tug forwards on the lead rope at the same time and tap the horse lightly and encouragingly on the croup using the whip. Horses should learn to respond to the smallest of signs.

Remember to practise leading from both sides.

EXERCISE 4: TURNS

Aim of the exercise

Turning to the outside is a great way to get a 'frozen' horse moving again. Turns also train body image, balance and communication between handler and horse, and are good for releasing tension.

Turns followed by halts in the labyrinth. Liberty is fully engaged and focused.

What you need to know

A charging horse can be brought under control by turning it to the inside or by using serpentines if there are no ground obstacles nearby. Always make sure to stay in front of your horse's shoulder. Don't turn from a standing start; lead the horse straight ahead first.

What you need

Properly fitting head-collar, rope with quick-release panic clips, whip. Ground obstacles made up of foam tubes or poles make this exercise easier.

How to approach the exercise

Signal the horse to walk on. After a few paces in a straight line, turn to the inside or outside. Use aids, i.e. by combining body posture, whip signals and tugs on the halter. Your body should point in the direction you want yourself and the horse to go. Keep the rope short. You can support a turn to the outside by pressing lightly on the horse's cheek. This helps the horse to bend better into the turn.

When practising turns, remember to lead from both sides.

EXERCISE 5: GOING BACKWARDS

Aim of the exercise

This is a basic exercise in initial training. It is important when training a horse to load and improves attentiveness and balance.

What you need to know

To be able to reverse or 'back', your yearling needs to shift more of its weight backwards. To do so, it needs to raise its head slightly, but should not lift it abruptly. The sequence of steps is fluid and diagonal. Tense, hectic backing up as a punishment is not what we are aiming for.

How to approach the exercise

Stand beside your yearling and, with the thick end of the whip or your fingertips, tap him on the chest above the point of his shoulder. Give a command ('back') and move towards him. If this is not enough, place your hand on his nose band or chest and push him backwards. If he shifts his weight even slightly, reward him with praise and a short break. Preparatory foot and leg work, such as stroking and gentle massage, can be helpful.

EXERCISE 6: BESIDE, BETWEEN AND OVER TARPAULINS

Aim of the exercise

This exercise builds trust, gets your horse used to different surfaces and teaches handler and horse to pay attention to each other's body language. It is also good preparation for travelling in a trailer.

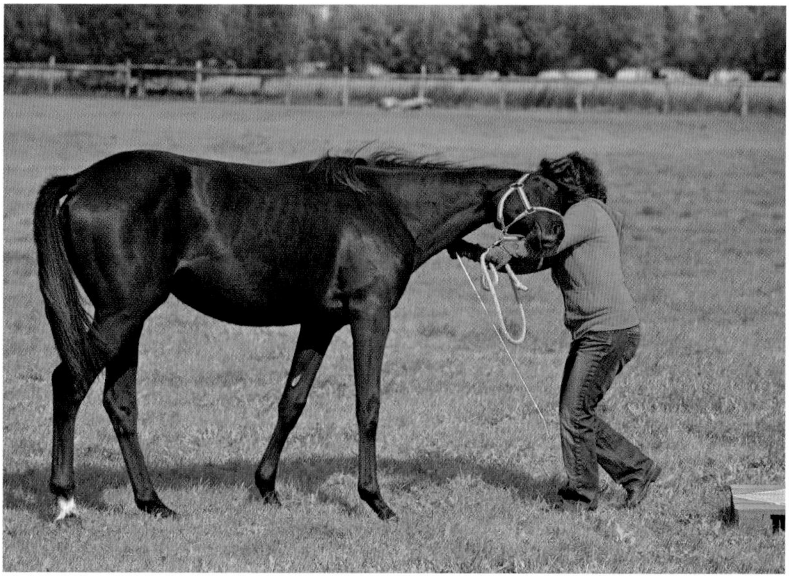

Most young horses don't enjoy going backwards. Universe gripes at first.

What you need to know

Some horses walk over the tarpaulin calmly at first, and only spook once they have done it a few times. So always stay to the front and side of the horse and take a few steps backwards if necessary. Some horses need only a day to master the steps described below. Most horses take several days or even weeks. Content yourself with small successes! Break the exercise down into as many small chunks as possible; this maximises the chances of your pupil being brave and succeeding at each step.

How to approach the exercise

Lay down two plastic tarpaulins in a V shape with a wide gap in between. Lead the horse past the tarpaulins on the outside, then turn around and repeat this step so that the horse has seen the tarpaulins with both eyes. Then change sides yourself and repeat this step again. If your horse is relaxed, lead it through the tarpaulins in both directions and from both sides. Next, walk over the tarpaulins yourself, without leading the horse over them. This gets

But in the end she lets herself be persuaded. A small step is enough and is rewarded with huge enthusiasm and the end of the exercise.

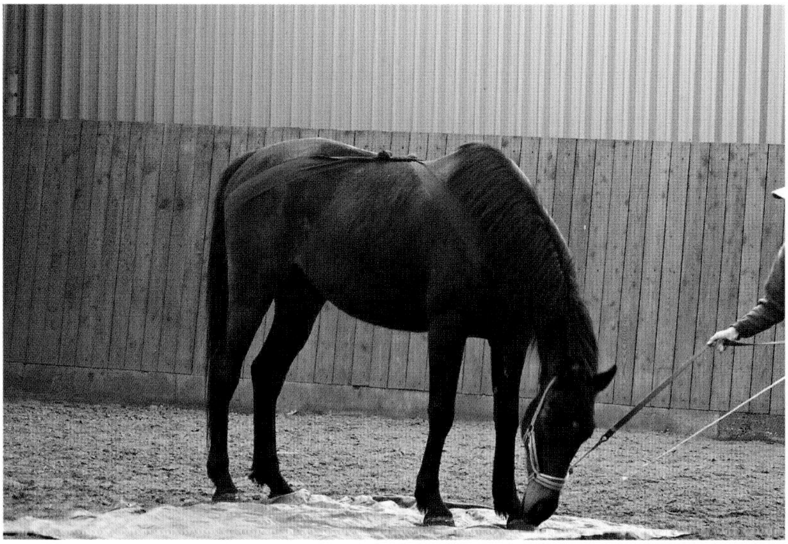

Hero is given a long rope as he stands on the tarpaulin, allowing him to sniff it properly. Note how well-balanced he is, with a 'leg at each corner'. Patient preliminary work in between the tarpaulins has helped to relax him.

it used to the noise they make. Now narrow the gap between the tarpaulins slightly.

Leading from the wider end of the V makes the obstacle seem more inviting. Gradually reduce the distance between the tarpaulins and continue to walk across them without leading the horse over them. Stay to the front and side of your horse: a horse that is calm at first may suddenly decide that it has had enough and try to get away.

In the next step, allow your horse to investigate the tarpaulins. It will smell them, paw at them and may even take a corner into its mouth. Young animals are inquisitive! Encourage it to stand on the tarpaulins and walk across them. Don't introduce the next step until your horse has mastered the previous step and has stayed relaxed throughout.

If your horse was brave at first but is increasingly showing reluctance, you have taken things too quickly and it is time to change the exercise or call it a day. If you start over again (from step one) after a few days' break, the horse's reluctance will usually have disappeared.

Universe enjoys stamping on this pallet, checking to see if it really is stable and safe.

EXERCISE 7: PEDESTAL

Aim of the exercise

This is a trust-building exercise which prepares your horse for trailer training.

What you need

A solid, stable pedestal such as a wooden pallet reinforced with floorboards, a whip, a lead rope and a properly fitting head-collar.

What you need to know

Always allow your horse to step down backwards from the pedestal and make sure it doesn't cross it in one big stride. Your horse should already know the command for going backwards.

How to approach the exercise

Position your young horse at a slight distance in front of the pedestal. Stand to the front and side of the horse. Allow the horse to smell the pedestal and to investigate it with a hoof.

This is enough for a first time. However, many horses are adventurous and inquisitive enough to step on the pedestal. If your horse does this, steady the pedestal with your foot. Allow the horse to stand for a while and then lead it backwards and down from the pedestal.

This yearling mare stands bravely on the pedestal. She can expect praise, a treat and a break from work!

GROUND WORK

Prepare thoroughly for your short training sessions. Arrange to work with at least one other handler and their young or adult horse. Think about which exercises will be useful and feasible, plan the order in which you want to do them, and build any obstacles before fetching your yearling. Get some feed dishes ready if necessary. Once you have made these preparations, you will be able to relax, concentrate and spend a successful 10 minutes or so working with your young horse. You will be surprised how quickly these short lessons start to bear fruit!

Feed calms and motivates.

Exploring ground obstacles on a long rope.

Anyone can go fast. Slow steps promote a sense of balance.

Concentrating on a curved line.

Touching and sniffing are allowed!

Training to load

Trailer training begins on a simple wooden bridge, without a trailer in sight. Sides made of jump stands and poles are used to simulate the enclosing walls of the trailer. Once the horse has learned to walk on to the bridge and step off backwards while staying relaxed, you can introduce the trailer. Ideally, you should be working in pairs. Remove the partition and leave the front door open.

Lead your horse around the trailer, incorporating about-turns, figures of eight and plenty of pauses.

Lead the horse up to the ramp and stroke down its legs with the whip while a helper holds out a dish of oats towards the horse. If the horse places a first hoof on the ramp, reward it, give it some feed and push it back off again. Repeat this process several times: hoof on the ramp, praise, feed, push back.

Body bandages and frequent stroking with the whip help the horse to feel safe and secure. Here, Hero stands on an open-sided bridge.

After patient preliminary work, Hero loads calmly. At every pause and in the trailer itself, the handler continues to stroke his legs with the whip.

If there is no pressure of time, you can repeat this exercise daily until the horse is standing fully in the trailer. Once the horse is standing calmly in the trailer, push it back out again gently before it decides to reverse out for itself. Always end the exercise on a positive note for the horse (and for yourself).

Now you can put the trailer partition back. At first, keep the partition wide open. Once the horse is standing relaxed inside the trailer with the partition in place, adjust the partition gradually until it can be locked.

To prepare the horse for the feel of being restricted by the breech bar, disconnect the bar and touch the horse with it, concentrating on the croup and hind legs in particular. Use the bar to push the horse forwards slightly.

Now reconnect the breech bar and calm the horse with soothing words, and by stroking or scratching it on the rump. If the horse tries to walk backwards, it will feel the restriction. It may spook or panic, possibly even trying to jump the partition. But with careful preparation such events are rare.

If the horse is relaxed, you can close the trailer door. The rope is pulled through the tie ring, the helper stands at the front in a safe position and the horse can be given feed to encourage it to chew and relax.

If the horse stays relaxed, it is time to start the engine and get the vehicle and trailer moving.

At each stage of this exercise, it is very helpful if the green young horse can watch how an experienced 'easy loader' deals with the trailer. Its first

Trailer trauma

Some horses load easily from the start. Others need two weeks of careful preparation before making their first trip. Once a horse has been traumatised by a bad experience, it usually takes several weeks to get it in a frame of mind to travel again.

short trips, as well as longer journeys later on, will be less stressful by the side of a calm old hand who has done it all before. But before this, make sure to practise loading with a second horse in the trailer. And when you practise, always lead from both sides.

The two-year-old on summer pasture

Management and feeding of two-year-olds

AGE
24 to 30 months

When the weather and grass growth allow, the two-year-olds can be turned out on to generous, securely fenced summer pastures from late April until late October. The ideal pasture is slightly hilly, with a wide range of plant species, non-poisonous hedgerows to nibble on and trees or field shelters to provide shade.

Turnout should be limited at first, to prevent digestive problems and excessive protein intake. After a week or so the two-year-olds can be given 24-hour turnout. In the first couple of weeks they are fed supplementary high-fibre forage such as straw or coarse hay from the previous year. Forage is especially important in spring and autumn and after long periods of rain, when the grass is abundant but low in crude fibre. In May and June, well-managed pastures should provide sufficient food for young horses. Supplementary feeds are needed only in very dry summer months and are available from a wide range of manufacturers. Fresh drinking water and a salt lick should always be provided, and mineral feeds and forage should be fed as required. Mineral licks have proved very useful, with experience showing that horses use them as and when they need to.

The horses' keepers are pleased that the labour-intensive stabling period is over, while the animals themselves express their pleasure at being granted their natural freedom by galloping exuberantly around the pasture, which should be large and well-fenced. As in previous years, the young horses are still in their familiar groups of friends or relatives. In the end, the scent of fresh pasture grass wins out over the fun of racing around and making and accepting friend requests. Freedom to run about with their peers helps young horses to mature in a healthy way. If they are being reared under natural conditions and have constant contact with people, they can now simply concentrate on growing.

Our mares live in a large, mixed-age mare herd. Stallions are better divided into separate male-only herds based on age.

If a stallion is to be presented for licensing, it will need to be fed a suitable energy-rich diet, which is not necessarily beneficial for its health (see p. 122).

The young mares stay together with their friends, aunts and mothers in a large, mixed-age herd.

If you plan to entrust your colt to a stallion trainer, find out as much as you can about how the yard treats young stallions. After all, you want to preserve the animal's psychological and physical health (see p. 56).

Stallion licensing

If you have ever attended one of the many stallion licensing events in the spring and autumn, you were probably surprised to see how 'ready' and mature the two and a half-year-old stallions looked in only their third autumn. But if you could see right through them – as a good judge of horse-flesh can – you would find an immature musculoskeletal system which is already having to support a lot of muscle. Often far too much muscle, in our opinion. There is a saying among experienced equestrians: 'An extra year as a foal, 10 extra years as a riding horse'. By this they mean that horses should be broken in not at three years old, but at four. The longer they are reared under natural conditions with plenty of exercise, and the steadier their growth rate, the healthier they will remain. And the later they are put into

Hopefully, the two-year-old stallions will still be able to enjoy the summer together with their peers.

work, in riding or as carriage horses, the longer you will be able to enjoy a healthy horse that is keen and ready to perform.

Pushing the young animal unnaturally, either by overfeeding concentrates or by introducing work and training early, is clearly harmful to its health. In our opinion, it is also unethical.

Like an adolescent during puberty, your two-year-old is not yet fully mature. It is still 'soft', both physically and mentally. It is not able to cope with high levels of work and has a tendency to overreact. Keep it occupied, give it an occasional easy task to make it think, and remember that it only has a short concentration span and is vulnerable to physical injury.

Walks, basic training exercises, moderate ground work, a few circus tricks or composure training exercises in the company of other horses, as well as easy exercises in preparation for breaking in, will feed the young horse's curiosity and strengthen its trust in you. Break exercises down into lots of individual steps and increase the degree of difficulty only once your horse is participating in a relaxed manner.

Stallions with licensing prospects often don't have the luxury of summer pastures any more. They are put to work at an early age in preparation for

Marked out for a life as geldings, these colts can continue their carefree existence, unlike some of their friends. Candidates for licensing are stabled as two-year-olds and will spend the summer indoors undergoing preparation.

First Flight Spirit was licensed at three and a half years old, so he was able to spend his second summer at pasture.

licensing at only two and a half years old. Their daily routine consists of stabling, daily lunge work, exercising in a horse walker, free jumping and often a fattening diet.

In 2009, the German Trakehner Association decided that stallions directly descended from a thoroughbred parent could be presented for licensing a year later because they generally mature later than full-bred warmbloods.

We believe that this rule should be extended to all other breeds and Associations.

LICENSING AT THREE AND A HALF YEARS OLD

We took one of our half-bred horses off pasture at two and a half years old to prepare him for stallion licensing. Even though his conformation was still very foal-like, he did everything we asked of him and enjoyed the generous feed ration. Realistically, however, he wouldn't have stood much of a chance in the stallion licensing compared with the much more ready-looking warmbloods. On learning of the change in the Trakehner Association's regulations, we broke off the young stallion's training immediately and returned him to his friends in the pasture. They had been apart for only a couple of weeks, so we took a chance on putting them back together even though the pecking order in the little herd had changed. You couldn't do this with every stallion, but this group had grown up together in a real spirit of camaraderie. So putting them back together was not a problem.

This stallion was therefore able to enjoy another whole season on pasture, play with his friends again and mature in a more natural way. First Flight Spirit was then licensed as a three and a half-year-old in the following autumn.

Young blood

Watching in amazement ...

... feeling good ...

... showing off ...

. . . playing . . .

. . . debating

A horse's teeth

Adult mares have 36 teeth, while stallions and geldings have an additional four canines, making a total of 40. Some mares can also have canines.

Horses have six incisors and 12 premolars or molars in both the upper and lower jaw.

Unlike human teeth, horses' teeth are worn down constantly by grinding and grow by around two to three millimetres each year. The canines are an exception, as they are used not for breaking down food but for fighting.

In foals, the incisors erupt through the gum around a week after birth. The incisors include the first (central), second (intermediate) and third (corner) incisors. The milk teeth or premolars erupt either at birth or within the first few weeks of life. There are six of these in both the upper and lower jaw.

Behind these, the grinding teeth or molars erupt between the age of one and four. The canines erupt after four to six years.

The central incisors are replaced at the age of two and a half, the second incisors at three and a half and the corner incisors at four to five years old.

The premolars are replaced at the age of two and a half to four years old.

The wolf teeth are an exception. Wolf teeth are vestigial premolars, remnants from the evolution of equine teeth. They are small, peg-like teeth without a large root. Not every horse has wolf teeth. And not all wolf teeth erupt through the gum, in which case they are described as 'blind' wolf teeth. Wolf teeth may cause pain during leading, lunging or riding with

Herbal remedies for teething troubles

For painful tooth eruption, halter sensitivity, biting or head tossing, natural healing practitioners give five drops of Chamomilla C 30 once a day for three days.

a bit, causing the horse to fight against the hand. We recommend that wolf teeth are removed by a specialist before a snaffle bit is worn for the first time.

Exercises for the two-year-old horse

You can of course continue to do all of the previous exercises at regular intervals. But a new aspect of the horse's training, intended partly to prepare for its future as a riding horse, is mouth work.

EXERCISE 1: MOUTH WORK

Aim of the exercise

Mouth work has positive benefits in preparation for worming, equine dentistry and wearing a snaffle bit, during tooth eruption, and in horses with a tendency to take everything into their mouths and chew it.

What you need to know

Many horses will pull their heads away if you try to touch their mouths. The mouth is an intimate area for horses, just as it is for humans, and it is particularly sensitive during teething. A horse's mouth is closely tied up with its emotional life.

How to approach the exercise

Content yourself with brief successes at first. With time and experience, your horse will come to enjoy mouth work and you will be able to extend the exercise.

Start by touching and massaging the area around the mouth and run a relaxed hand over the horse's nostrils, lips and chin. Trace around the insides of the nostrils and stretch the lips and nostrils a little. Slide a finger under the upper lip and rub the gums, or find another way into the horse's mouth.

Rub, touch and massage all of the areas that you and your horse find pleasant, including the soft palate, tongue and insides of the lips. Some horses are more relaxed afterwards, while others seem 'thoughtful' or preoccupied with their mouths.

Use a head-collar when starting mouth work, so you can hold the horse to you if it tries to move away.

From two-year-old to three-year-old – the third winter

Management and feeding

AGE
30 to 36 months

The transition from pasture to stabling requires careful preparation in order to avoid digestive disorders. A good option for a smooth transition in cold, damp weather is night-time stabling. If the weather is mild and dry, the horses can be provided with forage out at pasture. Here again, rations should be gradually increased. Energy-rich concentrates are also recommended.

Two-year-olds still need several hours of exercise outdoors every day, even in winter. Exercise areas must be large enough to allow all of the horses to let off steam, roll and engage in social contact.

Exercise helps the muscles and bones to develop properly and maintains a healthy spirit, heart and lungs.

TWO SUGGESTED RATIONS		
In kg/day for two-year-olds 30–36 months old, medium-sized breeds, weight approx. 450 BW according to Meyer/Coenen (2002). BW stands for body weight in kg.		
Hay, first cut	5 kg	2 kg
Wilted haylage		6 kg
Carrots		4 kg
Fodder beet	6 kg	
Oat grains, crushed	1 kg	1.5 kg
Supplementary feed for foals	1 kg	
Supplementary feed in addition to hay/oats		1 kg
There should also be good-quality straw from well-maintained bedding. A salt lick should always be provided.		
The daily rations given in the table should be divided into three portions.		

More ambitious exercises

Our two-year-olds already know all of the basic exercises from when they were foals and yearlings. So the foundations for their training have already been laid and they are as well-behaved and polite as they can be – provided they are not upset by external circumstances and would rather run away than stand and listen.

As soon as they are doing the basic exercises competently and staying relaxed, the degree of difficulty can be stepped up. Familiar processes can be supplemented with new and slightly more ambitious exercises. Say, for example, your young horse halts nicely in the stable aisle or arena and stands still for a few moments at your signal. Now you can practise this in a less familiar environment, such as on the way to the arena or indoor school, or in the school itself. Increasing the degree of difficulty gives both horse and handler repeated opportunities to confront and deal with mild stress. The young horse learns to calm itself or to be calmed by a handler. The handler learns not to be influenced by the horse's restlessness but to stay relaxed and to actively help the horse to relax, for example by stroking it with the whip.

New ground obstacles can be introduced, such as a bridge, tarpaulins or new arrangements of poles. Standing between poles or cavaletti, driving

In preparation for breaking in, Flightfever learns to walk under an arch of whips.

from the ground and exercises designed as gradual preparation for breaking in present other manageable challenges. A light girth and body bandages can be worn during ground work.

If your young horse suddenly starts to regress rather than learn new things, it is time for a break. Perhaps you were slightly over-ambitious or didn't express yourself clearly enough. Or perhaps your pupil's unwillingness was due to physical growth processes (teeth, hormones, body growth) or other factors (changes in the weather, herd dynamics, etc.). Leave it for a few days before starting over again. This will give you time to think and identify the last point at which your horse was still doing well and enjoying working with you. This is where you should pick things up again.

But even when progress is good, retracing your steps a little and reinforcing lessons already learned is a good way to maintain and strengthen your horse's curiosity and its confidence in itself and in you as its handler.

EXERCISE 1: TYING TO A FIXED OBJECT

Aim of the exercise

This is a basic exercise designed to promote self-discipline and impulse control, and to prepare for breaking in.

What you need to know

As a precondition for this exercise, your horse should stay relaxed when being prepared for tying up. The area around the tie point should be quiet. On different days and in different situations, your horse will behave differently during tying up. Relaxation cannot be forced, but self-control can be learned. In stressful situations, apply the techniques described on pages 47, 77 and 141. If necessary, go back a step or two in the exercise in order to avoid injury caused by the horse shooting backwards. Always tie your horse to a loop of easy-breaking string or baler twine. The tie point itself should be

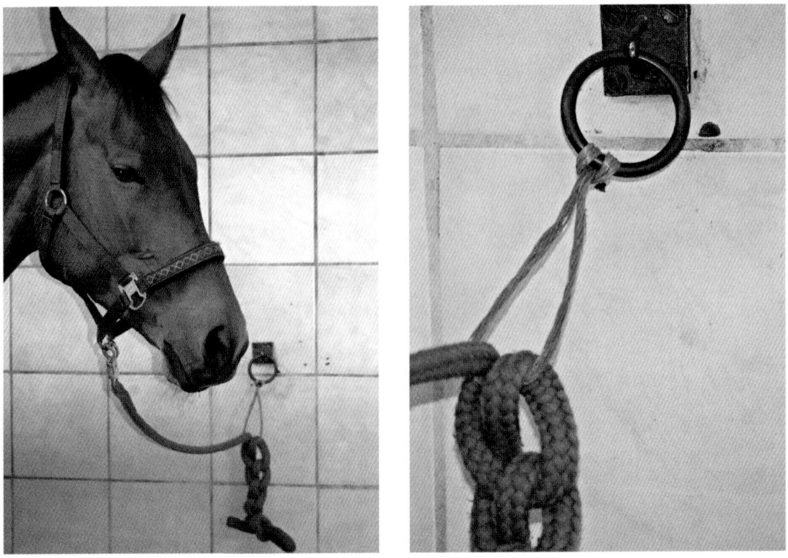

a) Sunset stands safe and relaxed, tied using baler twine and a tie ring (any string or twine should be easy-breaking). b) To tie securely, pass the rope through the loop of baler twine.

solid and secure. A metal ring cemented into the wall or a permanent hitching rail are good options for tying up. If you don't have a safe tying-up option, it is better to omit the exercise. Instead, work on getting your horse to stand quietly without being tied. Most riders have stories to tell about a horse that galloped around the pasture while still tied to a gate, leaving it forever spoiled for tying up. Foals, ponies and horses should never be tied to anything that is flimsy or unstable.

What you need

Properly fitting head-collar, rope, sturdy tie ring with baler twine. Never tie a horse using a knotted head-collar!

How to approach the exercise

Pass the lead rope through the loop of baler twine knotted to the tie ring. Use a quick release knot. Stroke and praise your horse if it stands quietly. Gradually extend the period of time for which the horse is tied up, ending the exercise promptly – as soon as you sense that the horse is becoming restless, if not before.

If your horse won't let itself be tied

Go back to the preparatory exercises (pp. 42 and 78). Practise standing freely as well. Never use coercion – tying up a panicking horse brings a high risk of injury (see p. 43).

EXERCISE 2: STANDING STILL

Aim of the exercise

This exercise allows you to observe your horse closely and to sense when it is about to move. It encourages patience on the part of both horse and handler.

What you need to know

As a precondition for this exercise, your horse should be familiar with the head-collar, rope and whip. It should halt when asked and be able to stand for a few moments (p. 103).

How to approach the exercise

Halt your horse and ask it to stand, as described on p. 108. Now walk all the way round the horse, maintaining physical contact the whole time. Horses are less tempted to move when you walk away behind them, and more tempted when you walk away in front. As soon as your horse understands the command to 'stand', put some distance between you. Start with half a metre, then increase the distance, always returning to the horse before it takes the first step of its own accord. Praise the horse and ask it to walk on. In some cases you may need a second person to halt the horse a few times while you walk away. If it walks after you, it is pushed back.

Always return to the horse before it moves, to make sure you have an opportunity to praise it.

Flightfever stands obediently. She enjoys a gentle rub between the eyes – a privilege accorded only to trusted handlers.

As she walks, Flightfever turns slightly to the left, towards the side she favours and the person she knows. This is one reason why leading from both sides is useful in order to improve balance.

EXERCISE 3: LEADING FROM BOTH SIDES

Aim of the exercise

This exercise promotes balance and straight lines, provides stability and relaxes the horse.

What you need to know

Leading from both sides is useful for young horses, so much so that you should include it in your daily exercises as often as possible! Nippy stallions and charging young horses can be kept at a distance, while skittish, speedy types can be encouraged more easily to concentrate on themselves. When leading, always make sure you can see your fellow handlers. If they fall too far behind, they have a 'driving' effect; if they walk too close to the horse, the animal can feel cramped.

What you need

A second person, a properly fitting head-collar, two lead ropes and two whips.

How to approach the exercise

Your horse should already be accustomed to being led from both sides. Attach the ropes to the lower rings on each side of the head-collar, stand in front of the horse's shoulder as normal and use the whips to point in the direction you want to go. Make sure to leave your horse plenty of room and stand as far out to the side as the lead rope allows. Agree with your fellow handler which route you are going to take and where you are going to halt. To slow the horse down, cross the whips far enough in front of its nose that it can actually see them. One handler may need to move towards its nose and give a gentle tugging signal on the rope. It also has a slowing down effect if the handlers are far enough in front of the horse that they can use their bodies, arms and whips to 'contain' it.

To halt, cross the whips in front of the horse. Beate strokes Flightfever as she stands still.

EXERCISE 4: UNDER AN ARCH OF WHIPS

Aim of the exercise

Your horse needs to get used to the sight of things above its field of vision – where it will see its rider later on – and to receiving visual stimuli from both sides. This exercise is also important for training to load, as it teaches the horse to lower its head.

What you need to know

This exercise needs to be built up step by step, and each step should be mastered with the horse remaining relaxed. Increase the degree of difficulty only once your horse has successfully mastered the previous step. The photo below shows the final step in the exercise.

Gala walks bravely under the arch of whips, which is now getting very narrow as the helpers stand closer together (see p. 137).

What you need

Two safe objects to stand on, three whips, two helpers, a head-collar and a lead rope.

How to approach the exercise

Step one: Lead the horse in between two helpers standing on raised objects. Make the distance between them wide enough to ensure that the horse stays relaxed.

Step two: The helpers raise their whips, holding them straight up, and you lead the horse in between them.

Step three: The whips are gradually brought together and crossed, and the distance is gradually reduced.

EXERCISE 5: BODY BANDAGE AND SURCINGLE

Aim of the exercise

As a step-by-step preparation for saddling and breaking in, this exercise gets your horse used to feeling restriction around its belly and to objects on its back, both at halt and on the move.

What you need to know

Many horses show different degrees of 'girthiness': fear of the girth or saddle caused by previous trauma. Even in relaxed situations, this fear can induce panic throughout a horse's life. Girthiness always has its roots in quick, sudden tightening of the girth without proper preparation. As flight animals, horses interpret this as an attack. This is why we need to proceed slowly and carefully.

What you need

Three simple elastic bandages from an equestrian supplies shop, a surcingle or lunging girth with rings, soft girth cover, head-collar, lead rope and whip.

How to approach the exercise

Start by getting your horse used to being touched in the belly area, by massaging it with the flat of your hand. Then gradually introduce the horse to body bandages. As a first step, wrap a bandage around its neck and chest and use this bandage to do your normal leading work. The next time you do the exercise, you can wrap a second bandage around the girth area and tie it to the one in front. Before tying the girth bandage with a quick release knot, use it to apply pressure to the belly area by lifting it up and letting it go again. Then lead your horse across some ground obstacles. The next time you do the exercise, you can wrap another bandage around its hindquarters. You

Body bandages make the young horse feel secure, encourage more conscious use of its hindquarters and get it used to a slight feeling of restriction.

a) Gala has still to get used to the bandage around her hindquarters. b) Hero is already quite at ease with it.

Hero is gradually being familiarised with the lunging girth and saddle cloth.

will need a second person to help you. Wrap a bandage around the horse's neck and tie one end of the second bandage to the first. With the aid of the second person, who stands holding the end of the bandage at a safe distance behind the horse, get your horse used to being touched on its hindquarters before tying the second end of the bandage to the neck bandage with a quick release knot.

Once your horse has digested these new impressions and stays relaxed, you can familiarise it with the saddle cloth and lunging girth. Start by stroking the horse all over with the saddle cloth before placing it on the horse's back. Place the girth carefully around its belly and apply light pressure to the belly area a few times by lifting the girth and letting it go again. Then do up the girth very loosely. To stop the girth from sliding backwards, use the third bandage like a breastplate, wrapping it around the horse's chest. Body bandages can also be used to give the horse a feeling of containment. Now lead your horse though its normal ground work and repeat the exercise a few times.

EXERCISE 6: DRIVING FROM THE GROUND

Aim of the exercise

To teach the horse to move independently without handlers leading from the side, and to tolerate aids on its body, i.e. where the leg of its future rider will lie. A good exercise for horses that spook at noises or movements behind them, it promotes balance and encourages them to shift weight to the hind-quarters. This is the most important exercise in the lead-up to breaking in and is good preparation for the lunge, double lunge and carriage work.

What you need to know

As a precondition for this exercise, your horse should be familiar with body bandages, saddle cloth and lunging girth. Put these on before starting the exercise.

What you need

Properly fitting head-collar, a helper with a lead rope, two 7-metre ropes or a double lunge, a neck ring, a chest rope or martingale, a whip, a lunging girth.

How to approach the exercise

Show a chest rope or neck ring to the horse and then place it around its neck. Loop the end of a long rope around the whip a few times and use this to stroke the horse. If the horse stays relaxed, attach the rope to the left side of the neck ring or chest rope and ask the horse to come towards you. The person holding the lengthening rope stands at a safe distance out to the side of the horse without initially touching it with the rope.

Both handlers go to the left side, so the horse can move to the right. Halt with a verbal command and a tug on the rope. Your horse already knows the command 'walk on'; what is new is the flick of the driving rein along its body. The handler at the head-collar helps by giving the horse familiar aids. Gradually, you can allow the rope to touch the horse's body while it is

Advanced driving from the ground without another person leading at the head-collar. This exercise is more useful than lunging to prepare the horse for being ridden later on.

moving, for example by making turns. If driving from the ground with a rope on the left side is successful, repeat the exercise with a rope on the right side. The person leading stays on the left side. If this step also works, and the horse stays relaxed, attach the second rope to the neck ring and do your normal leading work. The handler at the head-collar is still doing the leading and steering; but the signals to walk on, halt and go backwards are coming from the driving reins, used sensitively and with an emphasis on giving.

In the next step, the reins are attached to the side rings on the head-collar and passed through the rings on the lunging girth. Now the aids for turns can also be given from behind. Working in a fenced arena or indoor school, the person leading at the head-collar gradually becomes superfluous and your horse can be driven properly from the ground. A milestone in its training!

EXERCISE 7: WALKING ON THE LUNGE

Aim of the exercise

Step-by-step preparation for lunge work.

What you need to know

Once driving from the ground has become routine, lunging should not pose any problems. We recommend that two-year-old horses which are still growing should not be stressed with one-sided intensive lunge work. Unfortunately, intensive lunging of young horses is common practice when preparing for registration, performance tests or breaking in. The bones, tendons and ligaments of young horses are not yet stable or mature enough for fast-paced, narrow turns. If lunging is unavoidable when preparing for a performance test, limit the amount of lunge work and balance the stress on each side with frequent changes of rein. Gradually getting a horse used to side reins works best if

Here Gala is learning to walk around Marietta at an ever greater distance, as a first step towards later lunge work.

you attach them well in front of the vertical using elastic bandages. We also recommend that you teach the horse to work on a double lunge and with a lunging cavesson.

How to approach the exercise

During your normal ground work, take the lead rope into your outside hand and the whip into the hand closer to the horse. Move away from the horse, tapping it lightly with the whip at three points (behind the ear, on the withers and on the croup) to maintain the distance. Ask it to walk around you for a few steps. Then change sides. Include this exercise in your ground work from time to time.

Observe your horse closely in new situations, and radiate calm yourself.

Getting used to new situations

In the course of its life, your young horse will experience many new situations, some welcome and others less so. A helicopter landing nearby, a combine harvester coming towards you, a child riding on a toy car, etc. As you encounter and cope with each of these situations, your partnership will grow in cohesion, mutual trust and self-awareness. Ideally, you will deliberately expose yourselves to unfamiliar situations and practise calming your horse using the techniques described in this book. Choose appropriate challenges – not too small, but not too big either.

You can increase the difficulty of familiar exercises by doing them in an unfamiliar environment. Horses always associate an exercise with the overall process, including the external context. For example, if you have always groomed in the stall, your horse will need to learn from scratch how to be groomed elsewhere. If you have always practised halting along the side of the indoor school, halting in the middle of the school is a new exercise.

Gala's high head position betrays the tension she is feeling.

Until now, in the school or arena, you have always worked in the presence of a familiar older horse or even in a group of young horses. Now you could introduce a new horse and take away the familiar one. If this works, and your young horse stays relaxed, a next step in the exercise is to take the older horse away and to leave your young horse alone for a short while until a new horse comes in and joins it.

These steps are milestones on the road towards increasingly working alone with your horse.

Walking in hand is another way to vary familiar ground work exercises such as halting, walking on, turns, etc. Start with short routes with little traffic, and find a reliable horse to accompany you.

Calming techniques from your toolkit

Many of the exercises presented in this book can be used to calm your horse. But the fundamental principle is that you yourself should radiate calm authority and strict composure at all times.

Mildly stressful situations are therefore good for working on the aura that you project as a handler. Try to avoid being influenced by your young horse's alarm or excitement. Instead, train yourself to convey an attitude of 'nothing to worry about' through your body language.

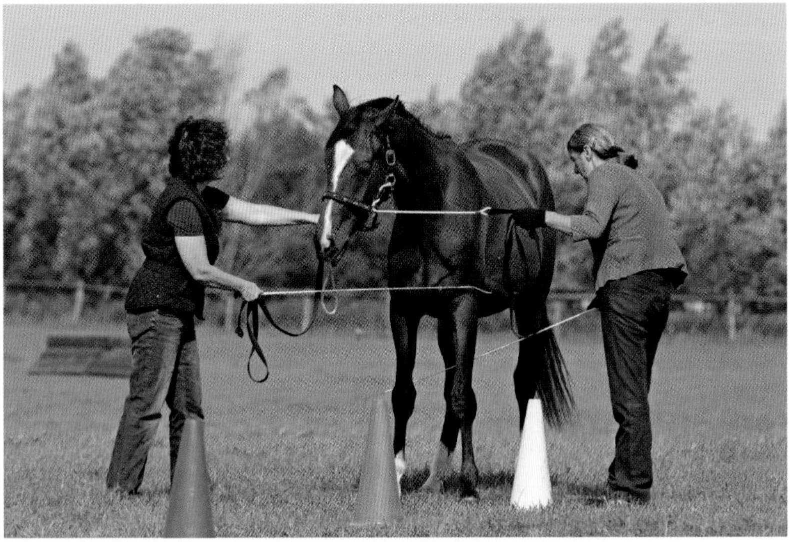

Stroking with the whip and leading from both sides are two ways to ease tension in the horse.

Breathe slowly and evenly, relax your shoulders and stay loose even if the horse is jumpy. At the same time, make your personal safety a priority. Sturdy footwear, gloves and maintaining a safe distance are a must.

By alternating the 'halting', 'standing' and 'walking on' exercises, you can distract your youngster and bring its focus back to you. Leading the horse through turns is also calming, as are leading from both sides, leading through ground obstacles and stroking with the whip. Body bandages, touching and the 'head lowering' exercise help to reduce physical tension. Feeding from a dish or by hand is distracting and relaxing, and encourages the horse to chew and breathe.

As a general rule, it is easier to assess the stress factors associated with a situation if you approach things systematically. Rather than work aimlessly, plan and design your work with the horse in such a way that neither of you becomes overstressed.

Feeding is calming because it encourages the horse to chew and breathe, and provides a short break.

EXERCISE 8: FREE JUMPING

Aim of the exercise

Prepares for registration and performance testing, encourages suppling, provides training variety, promotes body awareness, agility, obedience and muscle development.

What you need to know

Young horses should not be overstressed. One free jumping session per week is enough. Each time you do this exercise, end with a successful jump. Similar patterns of movement do occur in the wild but there the horse decides when to stop. Don't be too ambitious: end the exercise after six to eight jumps of up to 80 centimetres, even if you know your horse could jump higher. We do not believe in exhausting the jumping ability of horses as young as these by endlessly raising the pole. Although a two-year-old may be keen to jump, its muscles and bones are not designed for games like these that humans find entertaining. True horse lovers can recognise real jumping talent by the way the horse moves, even if it is only trotting over cavalettis.

What you need

For suppling, you need six jump stands for an oxer and a cross-pole; to prepare for a performance test, you need at least eight jump stands for a cross-pole, an upright and an oxer, plus jump wings or two more stands and a length of tape to mark the side of the jumping corridor. You also need at least six or eight poles plus two soft foam tubes to use as placing poles. Your two-year-old should be wearing a head-collar and tendon boots. It is useful to have a short rope that can be quickly passed through the head-collar ring when leading the horse. A second person carries the whip and a third person holds out a feed dish to the horse after the jump. It is important to familiarise the horse with the process in a stress-free way, so you will need three people at first. Later on, two of you will be enough.

How to approach the exercise

In your first free jumping sessions, the height jumped is not particularly important. The point is to familiarise the horse with the process and send a happy horse back to its stall. At first, build the cross-pole low enough that you

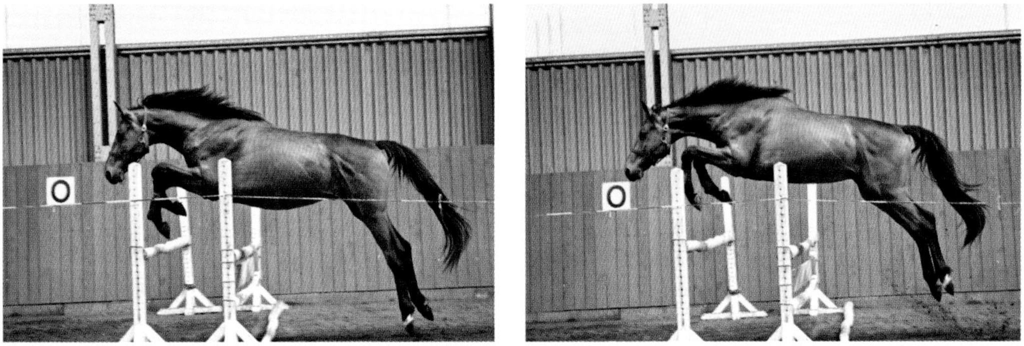

Even a low oxer trains coordination and strengthens the muscles!

can lead the horse over it calmly in walk. The third person holding the feed dish waits at the end of the corridor. If the horse is relaxed, it can trot over the low cross-pole by itself and is rewarded again at the end.

This exercise is enough for the first day. The next time, the horse will know what the jumping corridor means and you can convert the last jump into a small upright. Once the horse has successfully jumped the corridor a few times, you should stop. The next time, you can raise the cross-pole. This increases the challenge just a little each time. By the end, the horse will be jumping a cross-pole, an upright and an oxer, with each jump around the 70-centimetre mark. As your horse gets older, it can take more strain and is then allowed to jump higher.

Tip

The correct intervals between jumps are around 3.50 metres for an in-and-out, 7 to 7.20 metres for a canter jump and around 10.30 metres for two canter jumps. However, these intervals are only a guideline and should always be tailored to the individual horse and the training goal. To avoid having to fetch a tape measure every time, you can mark distances on the wall of the school.

EXERCISE 9: HEAD LOWERING

Aim of the exercise

To relax and calm the horse, encourage a relaxed posture and physical and mental composure.

What you need to know

Stressed horses raise their heads and tense up their muscles. Their breathing becomes shallow and ragged. To overcome this flight posture, it helps if your horse has learned to lower its head at a signal from you. Practise head lowering often, for example before going from a halt to a walk or while standing still. There are several signals you can give for head lowering, five of which are outlined below. Each of these signals is reinforced if you breathe out deeply and audibly when you use them. Try them out to see which one your horse understands best.

How to approach the exercise:

1. Using the whip, stroke from the horse's jaw, over its chest and right down to its hooves. Or:
2. Using the flat of your hand, stroke down from under the jaw to the shoulder. You can also massage the horse's neck as you move back up. Or:
3. Knead with the mane comb, working from the nape down to the withers. Or:
4. Tug downwards on the lead rope, as if you were trying to 'milk' the rope. Hunch your back slightly, and consciously relax your shoulders and neck. You can also rock the horse's head gently to and fro. Or:
5. Offer some feed from a dish held level with the horse's shoulder.

Flightfever lowers her head after some calming strokes with the whip.

EXERCISE 10: LEG WORK

Aim of the exercise

To prepare for the farrier, to loosen the muscles, to improve coordination of movements and to balance the horse.

What you need to know

A well-balanced horse stands like a table, with 'a leg at each corner'. But most young horses, and many trained ones as well, distribute their weight unevenly. Regular leg work helps to improve their balance.

How to approach the exercise

Start with yourself; stand as evenly balanced as possible with your knees slightly bent. Massage the horse's legs from the top down. Then lift up a hoof. Hold the leg firmly by the coronet and fetlock joint. Circle the whole leg round above the place where the hoof would stand. Don't pull the leg out to the side or towards you. Vary the size and direction of the circles, and the distance from the ground.

The circling should feel easy and effortless. Make sure that the movement comes from your whole body and continues in the horse's shoulder or

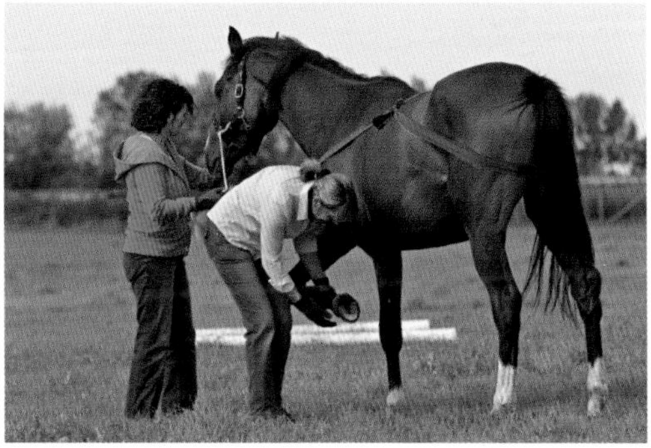

Leg work ties in with previous exercises and trains your young horse's sense of balance.

Circling the legs, massaging the limbs and tapping the hooves promotes body image and a stable standing position.

croup. At the end of the exercise, try to put the foot down gently and safely, tip of the hoof first. Work with all four legs and watch out for any differences from leg to leg.

EXERCISE 11: BALANCE EXERCISE

Aim of the exercise

This exercise improves your horse's balance before you even think about riding. Later on, it will be able to balance properly with a rider on board. The exercise trains coordination of the musculoskeletal system and, like riding half steps, teaches your horse to concentrate on you, pay attention to the smallest signs and carry weight on its hindquarters.

What you need to know

At this point we would like to re-emphasise how keen we are on ground work and encourage you to take it seriously as a training method for your young horse. We believe that ground work is the best tool to use if you want to have an easy-to-ride, cooperative horse under saddle later on.

Don't prolong the exercise for too long because it is very demanding in terms of concentration, for both yourself and your horse, and watch out for lapses of attention.

What you need

Ground obstacles, head-collar, rope, whip.

How to approach the exercise

Lead your horse as normal over various ground obstacles, quickly at first. Once it is familiar with the obstacles, slow the pace so that your horse is moving through and across the obstacles as slowly and deliberately as possible. In the next pass, halt the horse inside the obstacles and stroke it with the whip, slowing the pace even further. A more demanding option is to have

Stepping over a star and triangle, relaxed and with head lowered.

the horse step backwards out of the obstacles. Another challenging variation is the 'horse on a see-saw', in which, for example, you ask it to take three steps backwards and three forwards.

Practise from both sides and enjoy the interaction with your horse.

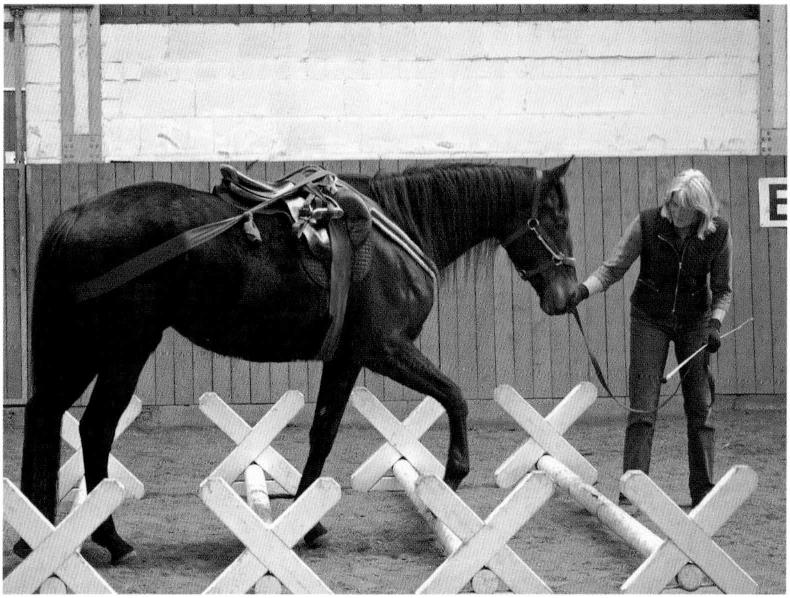

As flight animals, horses like to move quickly, including through obstacles. Taking only one or two steps forward at a time and then halting is a good learning experience for them.

WILLINGNESS AND COOPERATION

First Flight Spirit in a performance test

Royal Spacestar in a basic gait test

Raumpionier and Elmar Lesch

First Flight Spirit and Fabian Lipsky

Resources

EXERCISES FOR EVERY AGE					
Foal	*Weanling*	*Yearling*	*Two-year-old*	*Exercise*	*Page*
X	X	X	X	Making contact, touching	41, 81
X	X	X	X	Limiting freedom of movement	43
X	X	X	X	Wearing a head-collar	44
X	X	X	X	Giving a hoof, leg work	45, 86, 150
X	X	X	X	Leading for the first time	47
X	X	X	X	Preparing for tying up	48, 83
X	X	X	X	Mouth work, worming	49, 129
X	X	X	X	Ear tickling	50
X	X	X	X	Tickling the tail area	51
X	X	X	X	Circular TTouches	52
	X	X	X	Leading exercises	78
X	X	X	X	Stroking with the whip	82
	X	X	X	Stable aisle	84
		X	X	Halting, walking on	106
		X	X	Standing	108
		X	X	Walking on	109
		X	X	Turns	110
		X	X	Backwards	111
		X	X	Tarpaulins	112
		X	X	Pedestal	115
		X	X	Loading	118
			X	Tying to a fixed object	133
			X	Standing still	135
		X	X	Leading from both sides	136
			X	Under an arch of whips	137
			X	Body bandages, surcingle	139
			X	Driving from the ground	141
			X	Walking on the lunge	143
	X	X	X	New situations	144
			X	Free jumping	147
	X	X	X	Head lowering	149
			X	Balance exercise	151

Sources

Meyer-Coenen: Pferdefütterung, 4th edition, expanded and updated, Berlin/Vienna 2002

Möhlenbruch, Bottermann, Schwitte: Beruf Pferdewirt, Stuttgart 1999

Schäfer: Die Sprache des Pferdes, Stuttgart 1993

T.S. Stashak: Adam's Lahmheit bei Pferden, 4th edition, Alfeld/Hannover 1989

Tellington-Jones: Tellington-Training für Pferde, Stuttgart 2007

Voswinkel, Lena: Einfluß der Bewegungsaktivität auf Wachstums- und Ausdauerparameter beim Pferd, Dissertation. Christian-Albrechts-Universität, Kiel 2009

Further reading: Horse Riding and Management Series

Natural Medicines for Horses: Home Remedies and Natural Healing by Cornelia Wittek

Horse Behaviour: Interpreting the Horse's Body Language and Communication by Barbara Schöning

Useful addresses

Gestüt Heidekaten [Heidekaten Stud]
Dr. Hubertus and Beate Schmidtlein
23974 Heidekaten
www. gestuet-heidekaten.de

UNISONO
Harmony for rider and horse
Beate Schmidtlein
www.unisono-pferd.de

Ute Ochsenbauer
Animal health practitioner, author, horse trainer, riding instructor
www.uteochsenbauer.de

Index

5m
This practical and systematic book is an ideal guide for horse owners, breeders, riders and trainers, and also for equine science students, behaviourists and other equine para-professionals.
The focus of the book is on interpreting horse body language and signals to enhance understanding and the bond between human and horse. The book maintains a scientific approach throughout and discusses in detail the science behind horse behaviour – including the evolutionary basis of behaviour, the nervous system and behavioural development.
HORSE RIDING AND MANAGEMENT SERIES
Horse Behaviour
INTERPRETING BODY LANGUAGE AND COMMUNICATION
Barbara Schöning
5m Publishing
COMING SOON!
HORSE RIDING AND MANAGEMENT SERIES
The Horse Riding and Management Series is a practical and supportive advisory range of books for horse owners and people who have a passion for horses. The advice given is pragmatic, dependable and always expert. All facets of horse behaviour, husbandry, training and riding are covered to provide a fully rounded resource for equine enthusiasts.

5m
Natural Medicine for Horses is not only a perfect guide for horse owners but a great book for natural remedy enthusiasts too.
This books presents an essential A-Z reference guide to common natural medicinal recipes based on wild-growing and easily available flowers, fungi and herbs such as horsetail, chamomile and medicinal mushrooms. Explaining how they can be used and the potential benefits to your horse's health.
HORSE RIDING AND MANAGEMENT SERIES
Natural Medicine for Horses
HOME REMEDIES AND NATURAL HEALING
Cornelia Wittek
5m Publishing
COMING SOON!
HORSE RIDING AND MANAGEMENT SERIES
The Horse Riding and Management Series is a practical and supportive advisory range of books for horse owners and people who have a passion for horses. The advice given is pragmatic, dependable and always expert. All facets of horse behaviour, husbandry, training and riding are covered to provide a fully rounded resource for equine enthusiasts.